FEAR
OF A
DEAD
WHITE
PLANET

FEAR OF A DEAD WHITE PLANET

MORE WORLDS COLLECTIVE

Joseph Masco, Tim Choy, Jake Kosek, and M. Murphy

DUKE UNIVERSITY PRESS   DURHAM AND LONDON   2025

Project Editor: Livia Tenzer
Designed by Matthew Tauch
Typeset in Merlo Tx, Comma Base, and Cyrulik
by Copperline Book Services

Library of Congress Cataloging-in-Publication Data
Names: More Worlds Collective, author.
Title: Fear of a dead white planet / More Worlds Collective,
Joseph Masco, Timothy Choy, Jake Kosek, and M. Murphy.
Description: Durham : Duke University Press, 2025. | Includes
bibliographical references and index.
Identifiers: LCCN 2024057234 (print)
LCCN 2024057235 (ebook)
ISBN 9781478032106 (paperback)
ISBN 9781478028840 (hardcover)
ISBN 9781478061052 (ebook)
ISBN 9781478094333 (ebook other)
Subjects: LCSH: Human ecology and the humanities. |
Ecocriticism. | Environmentalism—Study and teaching—
Philosophy. | Nature—Effect of human beings on. | Global
environmental change. | Environmental degradation.
Classification: LCC GF22 .M674 2025 (print) | LCC GF22 (ebook) |
DDC 304.2—dc23/eng/20250210
LC record available at https://lccn.loc.gov/2024057234
LC ebook record available at https://lccn.loc.gov/2024057235

Cover art: Kathryn Cooper, *Smoke Screen*. Courtesy
of the artist.

Open Access publication made possible by generous
support from the Berkeley Research Impact Initiative
(BRII) sponsored by the UC Berkeley Library, the
University of California Davis Library, and the Division
of the Social Sciences at the University of Chicago.

CONTENTS

1   PART 0   INVITATION

4   PART 1   AGAINST THE ONE WORLD, FOR CONDITIONS

4   1.1   What Is a Planet?

11   1.2   What Is an Intergalactic Bummer Train?

15   1.3   What Is Environment?
                The Anthropocene's One World
                Leaving environment behind
                The geoengineering rush

29   1.4   Who, Where, What?
                Who are we, and who do we think you are?
                Where are we going?
                What is an X?

39   1.5   What Is a Core / What Are Worlds?

48   1.6   What Is a Species / What Is a Loss?

55   PART 2   WHO'S AFRAID OF A DEAD WHITE PLANET?

55   2.1   Situated Premise—Fear of a Dead White Planet
                The disaster is already here
                FDWP = F + D + W + P
                FDWP's illiberal and liberal modes
                Letting things die

78   2.2   Some Propositions
                There is no starting point, but we need to
                start somewhere
                The Charismatic Mega Concept (CMC) is a trap,
                but we still need concepts

Study will never be enough; orient to resolution

Don't trust your facts, but don't trust your objects either

Terraforming is not in the future; it is now

87 **PART 3 MIDDLES**

87 3.1 What Is a Middle?

88 3.2 What Is Land?

95 3.3 What Is a Lung?

99 3.4 What Is a Virus?

106 3.5 What Is Thinking?

115 **PART 4 TERRAFORMATICS**

115 4.1 The Resolve

124 4.2 Impossible Methods for Terraformatics Research Studies

Shifting microconditions

No one is alone

Terraforming for some can be terrabreaking for others

Take one step into something else

Welcome the contradictions

Unknowing objects

The future is in the past; time is part of the work

Accept impurity and commit to harm reduction

Conspiracy is not always wrong; consider making a conspiracy

Welcome no

Do what you are doing, maybe

Mistakes will be made

Pay attention to desire

Free causality from its metaphysical foundation and your study will follow!

Attend to matterings, not One World materiality

Still not sure how to start? Start with middles

138   **PART 5** **CONCLUSION AND FUTURE ASSESSMENT**

138   5.1 Welcome to the End

139   5.2 Gleaning Group III.5, Work Log 21220401,
               Tamalpais Archipelago, RSVTERRA9

143   ACKNOWLEDGMENTS / WORK HISTORY

157   REFERENCES

181   INDEX

PART 0
INVITATION

How do we study when the planet is on fire?

This question could go many ways. One could take it as a practical question. Which methods, tools, and concepts might help to solve the imminent problem of planetary emergency? One could hear it rhetorically: When so much is going wrong materially in the world, how could one possibly turn to studying as a course of action? Either of these could be followed in turn by interrogations of principles: What is *solving*? Who is the *we*? What is an *emergency* and for whom?

For our part, we assume that anything that big, that totalizing, that monumental a bummer—literally a planetary condition—implicates the organizations and practices of the university in a serious way. That a range of total Earth emergencies could develop so easily and garner such tepid responses in an age of abundant research and expert communication does not speak well for existing academic disciplines. They—their ways of slicing objects and methods, and their conventions for establishing authority and coherence—have been participants and executors in the current catastrophic state of things.

So, how to study when the planet is on fire? *Fear of a Dead White Planet* takes this question as a prompt for speculative thought and methods. We try to turn the question of *how* into an opening for stepping toward something different. This book is an experiment at building study in a joined-up way, the four of us, and maybe you too. It is an effort to lay some speculative groundwork for approaching the treatment of massive "environmental" problems differently. Our marking of "environmental" in quotes reflects one of our aims, which is to problematize the naturalized registration of drastically changed and changing lands, waters, and airs as "environmental" problems. We reject One Worldisms—that is, the assumption of a singular human species/planetary relation that can be characterized in universal terms—in favor of recognizing the world of many worlds: that is, the multitude of ways of living and being.[1] We become unfaithful to beloved concepts, like *planet*, *environment*, *land*, and *species*. We do so because, in the university, the fear of environmental chaos and efforts to stem it are often linked to insidious defense mechanisms protective of White Supremacy's foundations. We diagnose efforts to save the planet that maintain structures and forms of Whiteness as a syndrome we call Fear of a Dead White Planet. The singular planet it posits and venerates, far from needing saving, is a formulation of colonial Whiteness to be rejected.

Our ways into this problem take several shapes. We challenge the epistemic commitments and material practices of contemporary environmental research entangled in this pernicious syndrome. We join up with existing communities that already think against, and not just with, the university. We seek methods to recognize other relations and activate collective efforts to make less violent conditions. As an effort to build study in a joined-up way against the logics of White Supremacy, this book embraces the joints, not the smooth resolutions. Instead, we come to see study in all its unruliness as a vital part of worlding, that is, creating conditions and relations for coming together in less violent ways, even as we are in the middle of it all.

———

1 We take seriously the invitation of the Zapatistas to think this way and follow colleagues like Marisol de la Cadena in this approach. See Subcomandante Marcos (2022); and de la Cadena and Blaser (2018). In addition, see Povinelli (2001); de la Cadena (2015); Escobar (2020); and Omura, Otsuki, Satsuka and Morita (2018).

The result is a punctuated, difficult, even dizzying read, a book that is strident yet also repeatedly resets its questions, terms, and tone. This is purposeful. We are for hesitations, for taking things apart, for uncomfortably knitting relations together, for breaking some things and letting others die, as we are also for noticing what gets generated or lost in the differences between attempts to do study differently. We are after a form of convivial ongoing study across difference to build less hostile worlds, in the plural, here and now, again and again. We cultivate a kind of study as a collective modest terraforming amid the many fires. To begin, let's revisit the concept of the planet. Why on Earth do we think we even know what a planet is?

PART 1
AGAINST THE ONE WORLD, FOR CONDITIONS

1.1 WHAT IS A PLANET?

Out on the outer edge of our solar system, a set of small objects have been observed with strange orbits. The orbits, highly elliptical, make no sense; they should not warp the way they do, unless there is an enormous gravitational pull emanating from a source that no one on Earth has seen. Here is a beautiful problem of influence, where observation creates only a set of conjectures about the unseen, nonobject relation. Astronomers have created mathematical models of these orbits with unexpected eccentricities, repeatedly shifting the parameters of a simulated solar system over time to account for the observed behavior of these asteroids in the Kuiper Belt. The simulations are vast—they run on scales of billions of years and detail the movements of all known objects in the solar system—allowing researchers to test for influence or identify absence. A working theory is that there is a ninth planet, a huge Neptune-sized entity moving in an eccentric orbit around our sun, a thing big enough to pull smaller objects into odd elliptical trajectories.[1] This "Planet Nine" could be somewhere on an 8,000- to 15,000-year cycle around the sun, which might explain why no one currently alive on Earth has seen the thing.

From their Caltech offices in Southern California, astronomers Konstantin Batygin and Michael Brown did something remarkable in announcing their findings and the Planet Nine hypothesis: They invited the world to join in the search, welcoming everyone with a computer or a telescope to start looking for new objects, influences, and weird orbits and trajectories.[2] They gathered a global crowd to look up and out, searching the outer limits of astronomic perception for signs of Planet Nine. Very quickly the Planet Nine crowd began identifying new objects. The picture now emerging is not of a solar system run by the symmetry of perfect orbits, of moons orbiting planets and planets orbiting the sun—but an unruly universe of rogue objects, strange circuits, and a universe of nonsymmetrical possibility.

This is significant: The Planet Nine hypothesis has implications for how we come to understand the very terms of knowledge production. Despite repeated revolutions in computing, telescopes, space exploration, and satellite sensors, there has been no significant shift in the map of this solar system in over one hundred years. Now reflect on the possibility that an entire planet, one five to ten times bigger than Earth, could have been in our galactic midst all along. Or that the story of a solar system composed primarily of planets and predictable orbits missed an abundance of other forms, beings, and influences. What does this say about the apprehensive powers of academic research, or about our grasp of what makes a planet legible as such? It suggests we don't know jack about what is all around us. It suggests that the sense of *the planetary* noninnocently comes from particular histories and places and agendas. For instance, one of the key instruments mapping the solar system today is on Mount Graham in Arizona. Inaugurated as the Columbus Telescope, it was funded in part by the Vatican to mark the quincentenary of Columbus's arrival in the Americas in 1492. It not only nonconsensually occupies unceded San Carlos Apache Indigenous land but was named after

---

1 See Batygin and Brown (2016) for the initial proposal; and Batygin et al. (2019) for elaboration of the hypothesis.

2 For the citizen science project to look for Planet Nine, see Backyard Worlds: Planet 9, https://www.zooniverse.org/projects/marckuchner/backyard-worlds-planet-9.

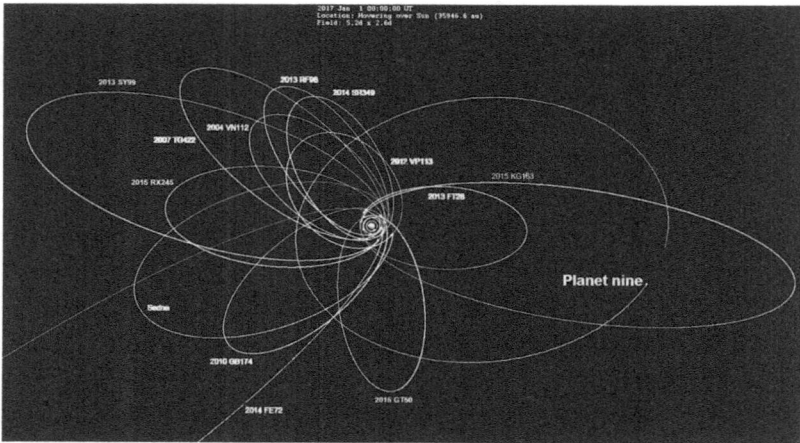

1.1 Possible location of Planet Nine within the Kuiper Belt.
Wikimedia Commons, 2017.

the famous colonizer who caused mass death in the Americas and "dis-covered" places by radically misrecognizing where he was.[3] How should we account for the legacies of this history in contemporary languages, disciplines, and perceptions? What else do we not know that we don't know about planetary conditions? And what other political, social, or material forces might be at stake?

Considered from Earth, the nonaligned objects in the Kuiper Belt, with their nonsymmetrical orbits and odd attractors, illustrate how bod-ies influence each other at a distance, creating degrees of sympathy over the vastness of time and space. From the vantage point of the Kuiper Belt, Earth might be a mere background to the eccentric pull of gravitational forces, nonsymmetrical behaviors, and planetary perturbers. And what would be the view right now from Planet Nine, or for that matter a Planet X, if we could register it? How limited would the earth science–based conception of *the planetary* as a system of systems seem? What might be the unseen or differently assembled intergalactic environment or, for

---

3   See Bainbridge (2020); and Castanha (2015). Tony Castanha discusses the tele-scope as a quintessential illustration of the ongoing force of the European "Doctrine of Discovery," a key aspect of the settler-colonial land grab. The Columbus Telescope was ultimately renamed the Large Binocular Telescope after fierce Indigenous land-back and environmental protests in the mid-1990s.

that matter, the unseen of any kind of environmental relation? Who is authorized to declare the planetary, and which institutions are so vast and self-assured to do so?

While scientists and social scientists alike are trained to think that direct observation is the way into every problem, for things that operate on a massive scale or that are radically distributed in time, one rarely sees anything but the fallout, the lag, the aftermath. But one can get good at looking for influences and interdependencies, searching for the effects within and between things and people, caring about the vibe, as well as the conditions and possibilities. Perturbations are an invitation to consider relationality, linking the forces that move an object to the anxious subject worried about how to survive current conditions.[4]

This book invites special attention to perturbations, looking for means of identifying influences, relations, uncertainties, glitches, unseen forces, tensions, and assumed objects. People have always used stars for navigation, creating their own systems and ways of knowing based on the night sky.[5] Migrations, underground railroads, and homecomings have been enabled by such knowledge. Many astrologies are based not on the map but on a theory of influence, assuming that planets in their movements affect not only tides but psyches, events, futures. One of the first moves of the modern Western academy was to reject this kind of thinking in favor of object-centered research—where an external observer was imagined not to be under the influence of the thing studied, let alone in a coconstitutive process with it. More than a century of big oil geopolitics and resource extraction has demonstrated the enormous power

---

4 Hannah Arendt (1958, 9) writes: "The human condition comprehends more than the conditions under which life has been given to man. Men are conditioned beings because everything they come in contact with turns immediately into a condition of their existence. The world in which the *vita activa* spends itself consists of things produced by human activities; but the things that owe their existence exclusively to men nevertheless constantly condition their human makers. In addition to the conditions under which life is given to man on earth, and partly out of them, men constantly create their own, self-made conditions, which, their human origin and their variability notwithstanding, possess the same conditioning power as natural things. Whatever touches or enters into a sustained relationship with human life immediately assumes the character of a condition of human existence. This is why men, no matter what they do, are always conditioned beings." See also Canguilhem (2001) on the related notion of milieu; as well as Turner (2013) and Benson (2020) on the concept of a surround.
5 See, for example, Harjo (2019) on Mvskoke sense of stars.

of this conceptual move, which has produced highly inequitable urbanized societies as well as a superheating earthly environment of megacities now facing spiraling calculations of imminent end-times. The "planet" emerges today as a strange figuration, wrapped up in the old Cold War space race, deployed by extractive industries and military states as a singular form susceptible to imagined managerial command and control efforts, even as the air, oceans, ice, and weather destabilize relations all around, leaving everyone to negotiate perturbations in their own ways.

What then is a planet? It is not a mere rock orbiting a star. A planet is a political and material figuration of incredible complexity, one that people encounter via influences working on vast and varied temporal and spatial scales. The figuration of planet increasingly is threaded through questions about habitability—both present and potential—and who a planet's inhabitants might be. Planetary consciousness is the production of decades, even centuries, of resource extraction, nuclear nationalism, racial capitalism, and the ambitions of entities (empires, big oil, counterterror, oligarchs) that seek to work on ultimate scale.[6] It is the figure of terra nullius, empty land for the taking.

The search for Planet Nine, for example, joins an energetic search today by NASA for exoplanets, or those entities outside our solar system that could potentially support human life. Some exoplanets orbit suns in predictable circuits, while others are on rogue trajectories racing through space. As the Earth heats up due to more than a century of industrial toxic releases accomplished by a hundred or so petrochemical companies consolidated in a few wealthy states, a new urgency characterizes the search for exoplanets, infusing this idea of planetarity with colonial codes of emergency and hope for a future escape route to other worlds.[7]

On Earth, the concept of the planetary is powerful today because for many it asserts a singular formation—the one planet—which has, can, might, or soon will not be able to support human life. The one planet we are trapped on. The one planet we might run to. The one planet is rendered a large object requiring large organizations and institutions (militaries, states, corporations) to manage its totalizing emergency. This formation

---

6   We thank Mary Louise Pratt (1992) for teaching us to think critically and historically about planetary consciousness.

7   See Political Economy Research Institute (2022) for a ranked list of the top one hundred entities contributing to greenhouse gas emissions. Most polluters on the list are fossil fuel companies but the US government is also in the top ten.

yokes the concept of planet relationally to a human species imaginary, producing a repetition machine of planet/human dualisms, overriding the multiple ways beings inhabit Earth and understand habitability itself.

Disciplinary ambitions around the planetary have been profound in the twenty-first-century North American university, revealing a set of perturbations in the logics of disciplinary thought and a new kind of conceptual land rush. In the past few decades, many geology departments have rebranded themselves as departments of planetary sciences, treating rocks on Earth as a model for landscapes on other planets, constituting the ultimate academic land grab as it stakes a claim on the entire universe of exoplanets. The geology of Utah stands in for Mars in many of the resulting accounts and models, while the top of sacred Mauna Kea in Hawaii stands in for the moon for NASA, creating an index on Earth projected outward onto other spheres.[8] Earthly habitability is thus the galactic gold standard even in its currently deformed state. The fierce debates around a golden marker and periodization of the Anthropocene—a purported geological epoch characterized by human-made changes—are similarly a mode of human planetary politics that seek to establish a singular planetary scale.[9] Much of the environmental humanities now risks accepting the singular planet as the object for its literary and historical reassessment, exchanging the multiplicity of worlds across languages for the One World currency of earth system science.[10] The planetarily scaled version of the problem of climate change has energized a fleet of techno-fixes obsessed with potentials for carbon removal and atmospheric engineering.[11] The result: the most thinkable response to a singular over-

---

8   For an assessment of exoplanets and concepts of home, see Messeri (2016); for expert assessments of the relation between geological and planetary sciences, see Vertesi (2015) and Olson (2018); and for a review of how Indigenous homelands are recruited into outer space research, see Hobart (2022).

9   For an overview of the official process, see Waters and Turner (2022). For important counterreadings, see Hecht (2018), as well as Whyte (2017).

10   The One World of the singular planet remains the master substrate and fuel for the environmental humanities even if particular research projects might fractionate it into different materials, processes, organisms, species, taxa, chemicals, and so forth.

11   This has a specific history in North America connected to post–World War II techno-optimism and settler logics of environmental improvement. For example, already in 1965 the National Science Foundation's Special Commission on Weather Modification declared that "weather and climate modification is passing from the speculative phase to the rational phase" (1965, 9).

heating planet is not a reassessment of disciplinary knowledge in the age of petrochemical capitalism but rather a charged push toward geoengineering.[12] Can we have a concept of the planet that doesn't relentlessly lead to the legitimation of a nonconsensual experimental reengineering of the entire Earth cast as a mode of rescue?

Observation and comparison are powerful tools but can also be used to discipline a One Worldism, an insistence that a singular version of the planet built out of North American imperial infrastructures of knowing and intervening offers the only map to the future. But how many unseen forces influence our bodies, imaginations, thinking? We know from examples around the world and in the past that universities do not have to work this way. How do we learn to ignore the perturbations always challenging our disciplines? Or shifting our bodies and beings around us? How do we learn not to see even the obvious? Knowledge production in the North American university has been historically constituted in support of extraction and empire, the precise engines of climate disruption and existential emergency today, yet disciplinary tool kits routinely erase this influence. The planetary is only one example of this. How are global heating, imperialism, racism, and mass displacement also made out of the missed or occluded insight, the rejected evidence, the silenced position? While intensifying climate disruption provoked the recent figuration of the Anthropocene, a project that focuses on naming a new geological era characterized by artificial material traces operating at planetary scale, geology's ultimate answer is not about dismantling fossil fuel capitalism but rather about creating a sharp timeline to mark the start of a new human-altered period that encompasses all the earth.[13]

What Planet Nine invites is a hesitation about such absolutes and humbleness about offering such a totalizing, nonnegotiable map. The search for a single planetwide pollution signal for the Anthropocene epoch simplifies rather than complexifies, elevating a single marker over

12 For an open letter by scientists advocating for geoengineering research, see Doherty et al. (2023). For a counterletter by scientists arguing for an international agreement preventing geoengineering research, see Solar Geoengineering Non-Use Agreement (2021). See National Academies of Sciences, Engineering and Medicine (2021) for an assessment of recent US investments in solar geoengineering projects.
13 For a sharp intellectual history of geology, see Rudwick (2014); and for a detailed account of how the US Department of the Interior became a global enterprise searching for oil and mineral profits, see Black (2018).

the vast number of perturbations that shaped and continue rippling through existing conditions. In this context of a One World objectifying planetary alarm, Planet Nine is a harbinger—not of planetary disasters to come but of planetary truisms to unsettle. Immense, unknown, and unfound, it begs suspicion of the planet's self-evidence and sufficiency as figuration and frame, and asks instead for more attention to the perturbations, for unlearning some core orderings, frames, and commitments of the university so we might recognize and build other worlds. This means acknowledging that environmental study is both constituting and being constituted by a concatenation of historically violent forces (racial capitalism, resource extraction, and nuclear nationalism to name but a few) that structure the university and inform modes of sanctioned inquiry.

To rethink the planetary is not easy today. It asks that we keep attuning to vast and varied violences that collaborate to create the ongoing destruction of living conditions on Earth—violences that also shape and inform the university. Denial of violence is a strong temptation, just as researching the concatenation of violence is exhausting. It is not easy to unlearn a mode of general inquiry that hides its origins and harms.

1.2   WHAT IS AN INTERGALACTIC BUMMER TRAIN?

Let us call this overwhelming problem space the Intergalactic Bummer Train. We're kidding, but we're also not; the phrase helps us with serious work. We name it to avow the affective challenge of going an immeasurable distance in a machinery of depletion, while reminding ourselves that we are riding on and through it collectively. As an alter-figuration, it helps us hold together the infrastructures (material, conceptual, affective) of the current emergencies while also resisting a One Worldism or being frozen by despair. It is a kitschy way of saying we are always already within the problem space, but we also have more agencies than we can ever know or admit. It's our way of staying with the trouble while keeping an eye on the opportunities of convivial connection in study. We imagine it with a soundtrack as well as heavy problem sets.[14]

---

14   Our sense of staying with the trouble is indebted to the noninnocent and genre-busting work of Donna Haraway (2016b).

Boarding this train means orienting ourselves to the challenge of study in a world of many worlds, where there are always other ways of being and knowing nearby. The train holds together uneasily. Some cars do not fit into each other's grooves; others resist being part of the train at all. Picture it not as a well-oiled high-tech machine, but as a precarious makeshift means to start coping with everyday violences that push everyone around but in different ways.

The train is intergalactic because in a world of many worlds, we are all surrounded by many ways of being and knowing, some easily aligned, some incommensurable. The world of many worlds is ever present. As we move intergalactically between cars and worlds to see who and what else is around, we know that each rider will not be welcome in every car. And some folks will not want to be linked in this way. Coming off the rails is part of the invitation to climb aboard. It is important to remember the mind-boggling universe of possibilities that surround.

The train is a bummer because the news is frequently not good, often overwhelming, and seems to be accelerating in one predetermined and terrifying direction. From air quality to storm fronts, from fires and floods to everyday violences across race and gender, from policing to aerial bombing to emerging resource wars and the increasingly desperate search for habitable planets elsewhere—our conditions can be bleak and hard to look at. The suffering is all too real, both maddeningly routine and intensifying. At the same time, the Intergalactic Bummer Train is an invitation to cut through this hardship in other joined-up ways, with respect, care, solidarity, and conviviality. The writing of this book, for us, has been a coming together in this way. Snacks, rest, tears, and bad jokes are part of the itinerary.

We figure our problem space as a train knowing that the train is a settler-colonial corporate engineering project in North America, one embedded in and empowering Indigenous elimination and buffalo genocide, anti-Blackness and immigrant labor, laying the tracks of racial capitalism.[15] Bumping along together on its bummer rails, we don't look away from the damning histories that made our ride. Consider how seamlessly the "golden spike" that completed the transcontinental railroad network was repurposed by geologists as the term for geological time markers and be-

---

15 See Hubbard (2014) on the interspecies violence of the intercontinental train, as well as White (2011). See Karuka (2019) and Schivelbusch (2014) on how the train changed settler notions of time and space and set up infrastructures for modern capital.

came central to current debates about the Anthropocene periodization.[16] When we hear the whistle, we think of Bong Joon Ho's 2013 science fiction film *Snowpiercer* in which a failed geoengineering experiment has pushed Earth into an inhospitable new ice age, leaving survivors trapped on a constantly rushing train that reproduces and maintains a vicious class system. It was the intercontinental train system that mobilized universal time and space, linking resource extraction to the emerging stock market—key infrastructures for what Manu Karuka calls "shareholder whiteness"—that is, the availability of intergenerational wealth organized through differential and promissory access to Whiteness.[17] The momentum of the train also offers up a sense of speed and nonresponsibility for the onrushing outside—an alienation from the land that it pushes through without regard. The Intergalactic Bummer Train calls out this conceit within the North American university. Refusing the straight line, going slow then fast then slow again, feeling all the attachments to our conditions, we practice riding while minding how so many parts of this train (and our attempts to think together) are caught up in these deadly histories. So we ask, what can a train become when it switches from intercontinental to intergalactic, and refuses to turn away from the bummer? How do we ride the train while also working to take it apart?

In this book, we will be making several stops on an itinerary of difficult situations to reconsider contemporary conditions marked in the university as "environmental."[18] Our goal is to attune to some perturbations that destabilize and challenge understandings of the planetary as a singular formation with seemingly overdetermined, targeted responses. On the Intergalactic Bummer Train no one can be a passive tourist: The ride is too bumpy, and we need your help in remaking and remaking again both the problem space and another kind of study.

We now invite you to join us on this Intergalactic Bummer Train,

---

16 See Zalasiewicz et al. (2019) for a compendium of technical studies for the golden spike of Anthropocene periodization, as well as McNeill and Engelke (2014) on the consumption patterns after 1945 that inform global warming.

17 For Manu Karuka (2019), shareholder Whiteness is tied to the possibilities for generational wealth produced by the linkages of the railroad, universal time, the telegraph and the stock market. Karuka also shows how these figurations constitute a countersovereignty to the multiple worlds already present in North America.

18 We put *environmental* in quotes here because it is a disciplinary concept we do not want to keep. For us, the environment is a concept to be taken apart, even rejected. We address this in detail below.

1.2 (*left*) The golden spike used to connect the Union Pacific and Central Pacific railways in 1869. Cantor Arts Museum, Stanford University.

1.3 (*above*) The golden spike geologists use to designate the Ediacaran Period. Global Stratotype Section and Point placed at Enorama Creek, Flinders Ranges, South Australia , in 2004. Wikimedia Commons.

where we will consider contemporary conditions and search for alternatives.[19] We seek new routes and different tracks, new collectives and different attunements. These include potential otherwises that might articulate different pasts and futures and that take the question of habitability seriously. Please remember, when joining the Intergalactic Bummer Train, the process is the point. Refusing the easy, heroic conceits currently offering escape from contemporary conditions—such as finding a new planet, inventing a geoengineering fix, or trusting in a utopian species-planet relation—it pursues a mode of joined-up work dedicated to recognizing both the violent conditions and the inadequate resources available for the necessary but always unfinished work of making more habitable worlds. Please also note that perturbations are tough, and conditions are unpredictable. You may nonetheless consider moving between cars; just listen for the whistle and mind the gaps. While the ride is rough and the turbulence is real, you are also in good company, and there

---

19. See Ra (2011) for a vibrant rejection of earthly conditions in favor of freer cosmic ones; and Szwed (2020).

is a lot to do—so hold on to your hat, and feel free to grab a snack and find a friend while we consider the *environment*.

1.3   WHAT IS ENVIRONMENT?

It is another day, and the news feed offers a feast of astounding environmental metrics. Arctic temperature is 25 degrees Celsius (77 degrees Fahrenheit) above average in the dark of winter when the next sunrise is still weeks away. At the bottom of the Mariana Trench, ten kilometers deep in one of the remotest areas of the ocean, marine scientists use a robot submarine to find microplastics in all deep sea creatures. In Paris, researchers find microplastics not only in wastewater and surface water but also, for the first time, as atmospheric fallout. In the UK, microplastics are found to reside deep in our lungs. In Japan, researchers find them in clouds. Meanwhile, the extinction rate is now one thousand times higher than background (i.e., the previous rate). In one year, the World Health Organization estimates, 7 million people died—one in eight total global deaths—as a result of air pollution exposure. In 2022, PFAS, also called forever chemicals, that are linked to decreased vaccine responses in children are found to be ubiquitous in umbilical cord blood. Coal mining and black lung disease are on the rise in Appalachia and in China. Storms everywhere are intensifying, as are megafires. Tiny particulates from increasingly routine megafires move across continental space to enter the lungs, then bloodstreams, of people and animals living far away. Oceans currents are speeding up due to increased winds from a warming atmosphere. The tenor of our times is streamed as crisis verging on apocalypse. The ongoing proliferation and accumulation of such measures offer up an optics of terrains, waters, airs, lives all massively out of order and seemingly getting worse.[20] What is happening?

---

20. On arctic temperatures, see Radio Canada International (2018); on microplastics at the lowest point in the ocean, in breast milk, and in infant lungs, see Carrington (2017, 2022a, 2022b); on microplastics in atmospheric fallout, see Dris et al. (2015); on microplastics in clouds, see Wang et al. (2023); on evaluations of the accelerating extinction rate, see De Vos et al. (2014); on global deaths from pollution, see World Health Organization (2014); on forever chemicals in umbilical cord blood, see Perkins (2022); on black lung disease rates on different sides of the planet, see Wang et al. (2024) and Kelly (2023); on intensifying megafires, see Trapp et al. (2007); on accelerating ocean currents, see Hu et al. (2020).

What footing can a critical researcher find in this turbulent stream of altered conditions that are at once so vast (operating globally) and so particular (identifiable at a molecular level within individuals)? Is there a general condition underlying the complex processes linking life forms and their possibilities for survival in a milieu of shifting land, sea, microbes, air, and ice?[21] What bundle of destructive forces are working at the scale of the planet but also concentrate specific trajectories of harm? What *longue durée* injuries amplify in magnitude with each passing year or are still to come? It is dizzying to connect data, affects, and imaginations under such destabilizing and unpredictable conditions.

Moreover, how can one study and respond to massive and ongoing environmental world-breaking in a way that doesn't reproduce that violence? What politics and methods of study can struggle within the conditions of impossibly hostile worlds? How might study be oriented toward becoming part of better worlds?

The obvious domain of such study is the *environment*, but we would like to start by demarcating the *environment*, that concept, as itself non-innocent and troubled. This is no small matter: Where you begin has everything to do with where you go, who you go there with, and how you get there. The term *environment* is attributed to Herbert Spencer, the scientific racist who also coined the phrase "survival of the fittest" in the nineteenth century, as a word for external circumstances, thereby implying a separation between oneself and outside conditions, and a natural justification for violent outcomes.[22] *Environment* smuggles in a separation of things (humans and nature, sociology and biology) that extends through the North American university.

This separation installs a set of forced choices between infernal alternatives.[23] If you start with environmental science, then the dots are

---

21   For ongoing reassessments of the concept of life via twenty-first-century scientific work, see Helmreich et al. (2022) and Lorimer (2020).

22   On the "infernal alternatives" of capitalism, see Pignarre and Stengers (2011), as well as Stengers (2015).

23   On this original formulation of the environment concept, see Spencer (1864, 414) and Pearce (2010); and for detailed genealogical studies of the term, see Warde et al. (2018), as well as Benson (2020). See Morton (2013) for a theorization of global warming as a hyperobject, something that dwarfs human sensibilities.

not connected to political conditions;[24] instead the research is likely reproducing the dangerous habits that it seeks to merely count, datify, and manage. If you find yourself in a course celebrating liberal environmentalism, then colonial conceits are quietly baked into environmentalism itself as a savior project (we at the university are here to save you). If you begin with environmental justice, then the materiality of both slow and quick racist killing is already what you are coping with and trying to dismantle, but the university may well categorize your work as unscholarly. The university may also demand neutrality in the face of massive death campaigns, from aerial bombing to forced migrations to genocidal projects that seek new land and resources for imperial designs—leaving the environment as simply a container for ongoing violence.

The *environment* as a problem space could be about understanding and dismantling uneven exposures to death that have been built into the very possibilities of attempting to live. Where you start could be about calling out and stopping practices that materially and epistemically make the breakability of land, airs, waters, and being itself as their method of aggrandizement. If we approach destabilizing earthly conditions as perturbations to our fields of knowledge (and not just out there in the environment), invitations to think differently abound. Influences matter to how we are studying and with whom and what.

The Anthropocene's One World

An unusually animated struggle over how to reframe the domain of the environment within the contemporary university has been organized around defining the Anthropocene. In 2009, geologists issued a global call to look for the clearest indicator of human alteration in the strata of the earth, the specific artificial planetary signature of a human impact that could anchor a new geological epoch called the Anthropo-

24 *Conditions* is a keyword in this book. By *conditions* we mean the always already altered material, imaginative, and affective dimensions that inform any given lived moment. We say much more about this later, but the important point is that conditions are processual and always in motion, and not confined to a sense of the natural as opposed to conventional understandings of the environment. Conditions are constituting modes of relationality and are always being influenced and organized by materials, actions, potentials, and rebounding agencies.

cene.[25] The resulting wide-ranging consideration of earthly states provoked a frenzied response across the humanities and social sciences, mobilizing scholars and disciplinary logics to debate the temporality, scale, terms, and the potential material planetarily distributed as signatures that could support such a designation (should it be carbon dioxide, plastic, plutonium, or maybe chicken bones, what about concrete or perhaps synthetic chemicals, and does it start with Columbus's trip to the new world, the invention of agriculture, the plantation system, the steam engine, or the atomic bomb?). This proliferation indicates how an incited multitude are trying to turn their skill sets to a sense of planetary environmental emergency, even as it becomes preconfigured by geologists as solely about periodization.

We find it helpful to consider the debate over the Anthropocene as both serious and diagnostic in relationship to the problem of the environment. Though newly coined in the early twenty-first century, the term already has (as of November 2024) over 511,000 scholarly citations and gains more by the minute. Its citation rise mirrors the classic hockey-stick line graph of sudden intensive increase of planetary temperature that characterizes global warming.[26]

The interdisciplinary tussle over the Anthropocene reveals a widespread awareness of destabilizing environments, starting with climate change and expanding into other planetary-scaled metrics, underscoring a deepening worry about human and nonhuman relations in the aftermath of generations of prioritizing petrochemical capitalism and nuclear nationalism over life.

It also reveals a stark problem in our disciplinary methodologies, a lacuna the size of the planet, in fact. For how is it that the vast modes of inquiry across the natural sciences, humanities, and social sciences could allow such a total planetary environmental condition to emerge as a new surprising frame? How is it possible that with such highly organized assemblages of disciplines, laboratories, global surveillance systems, social

---

25   See Crutzen and Stoermer (2000) for the original formulation; Bonneuil and Fressoz (2017) for an astute overview of the debate; and Zalasiewicz et al. (2019) for the technical findings of the Anthropocene Working Group as part of the official geological science inquiry.
26   See Osman et al. (2021) for updated planetary temperature graphs, and the Google Books Ngram Viewer (https://books.google.com/ngrams/) for the latest citation infographic on the Anthropocene.

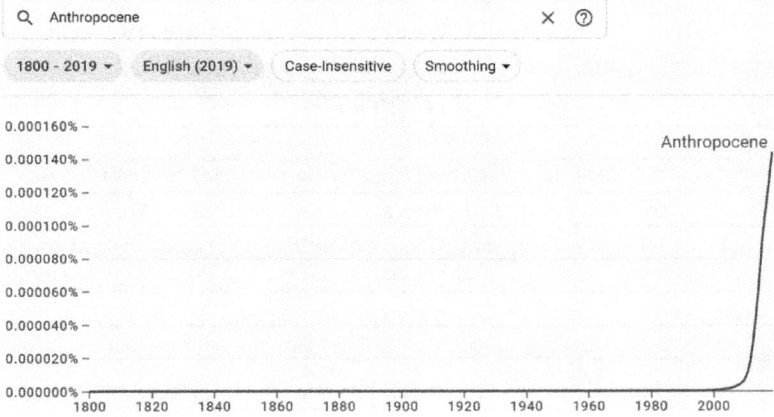

1.4 Google Books *n*-gram chart showing the use of "Anthropocene" in published sources from 1800 to June 2024. Courtesy of the authors.

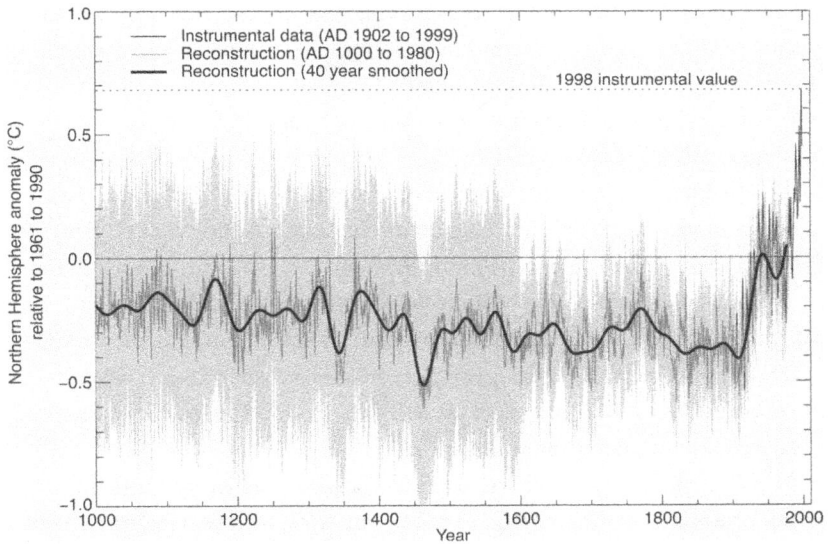

1.5 Northern Hemisphere temperature chart from 1000 to 2000 CE, known as the "hockey stick." From Intergovernmental Panel on Climate Change (2001).

theories, optics, and computational prowess—not to mention security agencies, international organizations, corporations, and NGOs—that a problem literally the size of the entire planet could emerge so prolifically, so late in the game? It might just reveal that academic (i.e., professional and professorial) processes of research and ideation install an infernal set of problems into the very grammars and apparatuses we rely upon for producing knowledge, expertise, and solutions. The result is a formatting of the university into generating false futures. These futures are not simply narrowly imagined. They are following tracks, routed by the university's ongoing enablement of profit built from death, as well as the quietisms and cynicisms invited by the propertizing, as well as careerification, of authorship and thought.

Monumentally scaled violence is not new. A better question is where to start. In the trending topics of academic research, a climate emergency can seem to be eclipsed by a war on terror, or a worldwide pandemic, or repeated global financial crises, and so on. Though there is an abundance of writing that shows how these structures of massive violence are linked and recursive, at the same time the disciplines of the university are constantly cutting these formations of violence up, presenting them as seemingly discrete. How do the very instruments of measuring and marking violent earthly transformations carry and enable the conditions of that violence itself?

This question cannot be answered by achieving a finer resolution or a better quantity of evidence or metrics, but instead indexes the disturbing ways that research can be complicit with the very problems it studies.[27] Nor will the problem be simply fixed through a holistic integration of the existing disciplines. In other words, when a problem literally the size of the entire planet emerges epistemically, we should ask: What does it reveal about existing forms of knowledge production, or about the way researchers ask questions and think?

What should we make of the fact that the satellite systems that are vital for understanding changes in the processes of the earth (from melting ice caps to rising ocean levels) are embedded in Cold War military systems of surveillance and nuclear war? That financial mechanisms are

---

27 For key theorizations of the relationship between disciplinary knowledge and violent conditions, see Fanon (2008); Du Bois ([1903] 2009); Deloria (1988); Judy (2006); and McKittrick (2021).

born through the violence of slavery and plantation economics? That scientific practices of observation and taxonomy are derived from imperial eyes? That statistics derives from eugenics? That public health derives from policing? That the concept of nature emerges via Indigenous dispossession and erasure? That data capture is yielded from surveillance practices built through segregation? What legacies of petrochemical industrial extraction, colonial management, racial labor, warfare, and imperial land grabs are embedded within the university, or the concept of the Anthropocene, or the earth system, or within our notion of an ecological relation or the solution of geoengineering? And so on.[28]

Scholars have recounted all these histories and more. But the serious work remains of addressing how the continuing techniques, capacities, and concepts in the North American university are ongoing second-order manifestations of the grammars of White Supremacy, both directly and indirectly doing destructive work in the present. In geology, ways of knowing and sensing derive in large part from the colonial histories of extracting oil from fossil remains in a profitable manner, a history that also brackets out other sensitivities as irrelevant noise and enables vast practices of wastelanding and elimination.[29] The way researchers come to know has a history, and a politics, and is not merely derived from the observation of matter.

Even when the planetary view offers to tame the cognitive disequilibrium triggered by the streaming news of environmental end-times, it cramps when faced with other streams of numbers, images, grief, and experience outside the purview of the environmental frame. The number of displaced persons, calculated at a mind-snapping 114 million in 2023, sets a new record. We can read tallies of migrants lost at sea (5,548 in 2014, growing to 8,600 in 2023), or track the rising stock prices of the boom-

---

28   On the dual use of satellites and computers for both earth science and warfare, see Doel (2003); Edwards (1996, 2010); and Masco (2006, 2014). On plantation economies and modern finance, see Burnard and Garrigus (2016) and Hudson (2019). On imperial eyes and objectivity, see Pratt (1992) and Haraway (1988). On statistics and eugenics, see Murphy (2017b). On public health and policing, see Ahuja (2016). On colonial ideas of nature, see Cronon (1995); Smith (2012); and McClintock (1995). On the origins of surveillance as racist policing, see Browne (2015).

29   See Bond (2022) for an account of how the devastations of fossil fuel extraction inform the concept of the environment, as well as Gomez-Barris (2017); see also Voyles (2015), for an important theorization of wastelanding in extraction zones.

ing industry in immigrant detention and biometric policing.[30] We witness the horrific death counts of genocide and war growing daily, as we also chart the profits from bullets and bombs and proliferating robotic killing machines on the stock market.[31] Unevenly distributed grinds of survival in hostile worlds, all-too-patterned destructions, habitual media alibis—these lie outside the environment as a research category, even as they should not.

Leaving environment behind

We struggle deeply with the concept of the environment and its disciplining habits, even as we engage its frame and recruitments. Among the four of us, some want to abandon it completely, believing that it is too freighted with imperial assumptions. Others hold that it has had, and has always held, other forms and relations within it, and that though the environment concept carries violence in its modern usage, it has always been, and can continue to be, a changing and contested form. The task therefore can sometimes be to embrace the term and transform it. Either way, we agree that for our purposes in this book, the term is too laden and limited to hold the place and do the work that needs to be done. We lean instead on the word *conditions*. By this we mean the always already altered material, social, imaginative, and affective dimensions that inform any given lived moment. We will say more about this later, but the key point is that conditions are not confined to old domainings of *nature* that

---

30   For the heart-rending counts of displaced persons, we look to the work of the United Nations High Commission on Refugees, UNHCR (2023), and we look to the Missing Migrants Project (2023), https://missingmigrants.iom.int/. The project documented over fifty thousand lives lost between 2014 and 2022.

31   The United States has been the world's biggest arms dealer for decades, providing 39 percent of all armaments between 2017 and 2021, including major support for Israel and Saudi Arabia; see Wezeman et al. (2022). This creates exemplary moments of planetary gaslighting: for example, in May 2024, President Biden's administration publicly announced an effort to restrict Israeli aerial bombardments of Gaza by theatrically withholding an order of two-thousand-pound bombs while simultaneously authorizing a separate $1 billion in new military support to the war machine of Prime Minister Benjamin Netanyahu; see Seligman 2024. This effort to calibrate, rather than prevent, mass death, occurred even as the International Criminal Court contemplated genocide proceedings against the state of Israel, exemplifying the formation of Fear of a Dead White Planet (FDWP) discussed in part 2.

subtract the social, a division which haunts conventional understandings of the *environment*.

Remember, the modern concept of environment paradoxically offers up objects as somehow separate from the subjects and socialites that are always fully within. This position of separation is an imperial fantasy of the empirical, where the reasoned observer can distance themselves from the object that is being observed, remade, and mastered. In Western thought, Francis Bacon and others dreamed of an empiricism that separated the observer from the object as a means of knowing it differently and perfectly, allowing for a mastery over nature conceived as a return to Eden, the garden in the Bible's origin story in which the first man, Adam, is granted dominion over and names all of creation. Scientific mastery is then envisioned as a type of divine Mini-Me of the Christian God, with its omnipotent sovereign power, achieved through rational dominion over an external world. Imagined here is a prototype for the geoengineer, the one that intervenes in all of nature through his expert knowledge of it. In Bacon's words, empiricism would both "let the human race recover that right over nature which belongs to it by divine bequest" and "establish the power and dominion of the Human race over the entire universe."[32] Baconian objects, Cartesian mechanisms, Newtonian atomisms, among many other moves, were part of European imperial rendering of beings into objects, lands into property, and relations into individuals under White heteropatriarchy. All this and more haunts the environment as a domaining of an existential political problem. It informs the long-sought power to control the weather, to remake geology, reorganize the biosphere, and colonize other worlds.[33]

The expert position of knowing and saving the environment delineates and twists forms of love and politics. Environmentalists easily become those who care about things outside themselves. Nowhere are these inherited distinct domains more clearly delineated and policed than in the university. There have of course been many alternative terms to *environment* within the genre; *nature*, *landscape*, and *ecology* are three of the

---

32   See Bacon (2009, 67) for more on this conception of a divine right over nature that simultaneously naturalizes conquest and social hierarchy.

33   See Fleming (2011) on the history of weather control efforts; Hamilton (2013) on geoengineering concepts; and Kaufman (2012) on US schemes to reengineer mountains and harbors with nuclear explosives.

most significant. But *nature* is too big, and in the Western tradition puts humans outside of it. *Landscape* is too representational and aesthetic, cultivated through bourgeois sensibilities. Even *ecology*'s foundational and radical interconnection of the world becomes delimited by its cybernetic birth. As a vision of objects interconnected in circuit diagrams, it is also disciplined by the university domains and its radical potential deeply usurped within the business logic of ecosystem services, digital and marketing ecosystems, the corporate or battlefield ecologies, and so on.[34] The *earth system* too has become domained within the divisions of sciences, protected by One World modes and planetary engineering ways of seeing, reproducing a society that would make Bacon proud.[35]

More than this, the concept of environment in the university rarely attends to the weaponization of matter into hostile exposures. It rarely is interested in the installation of public benches people cannot sleep on, or the uses of the deserts to kill and deter migrants, or the psychic hostility of White learning spaces.[36] Worse still, injuries by built material conditions are often arranged so that the exposed are made to appear responsible for the violence that they must endure. People are required to risk harm and death by the acts of working, escaping, living, and trying to survive, acts they must participate in and are exposed through. The concept of environment too often appears neutral to these violent relations built into and enacted through our material conditions. And the position of the individual separate from the environment helps render that individual innocent of the consequences of their acts. *Environment* is laden

---

34   For a foundational theorization of ecology, see Odum and Kuenzler (1963). For a detailed history of the term, see Golley (1993); and for assessments of its political impact, see Martin (2018) and DeLoughrey (2013). For a broader history of the influence of cybernetics across academic disciplines, see Kline (2015), as well as Halpern (2015).
35   The earth system concept refers to interlocking, interacting dynamics across land, ocean, atmosphere, polar ice, and geology. The concept was formalized in 1983, the multidecade achievement of Cold War environmental science and the broad conceptual embrace of cybernetics. For an overview of the emergence of the earth system concept, see Steffen et al. (2020); and on the related development of climate models, Edwards (2010). For an example of the kind of future projections that earth system science can offer, see Steffen et al. (2018) and McKay, Stall, and Abrams (2022).
36   On the weaponization of urban design, see Davis (1990); Soja (2013); and Weizman (2017). On the weaponization of desert landscapes, see Gokee et al. (2020) and De Leon (2020). For theorizations of racialized matter and education, see Wynter (2003); Deloria (2018); and Jackson (2020).

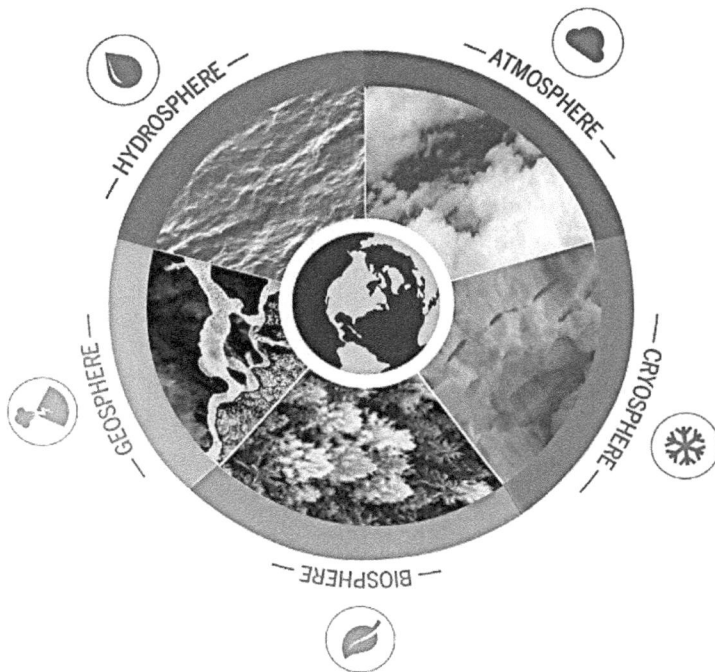

1.6 Aspects of the Earth system, 2024. From My NASA Data, https://mynasadata
.larc.nasa.gov/basic-page/5e-earth-system-story-map-collection-lesson-plans.

with these erasures and disconnections, and so we feel compelled to look for better terms.

Given the breadth of everyday violence, it becomes all too clear that reckoning with concepts of environment within the university cannot be fixed with apologies for compromised origins or the calling out of bad actors as founding fathers. The violences of disciplinary knowledge are not like upstream pollutants in the historical stream of knowledge, to be filtered, canceled, or remediated for safer drinking in the here and now; better to liken them to the carving of channels that desiccate some lands and lives while irrigating others. (Time to change channels!) In their very practices of discernment, the disciplines reproduce the world-shaping projects that installed them; they can't help but tame scales, histories, re-lations, and responsibilities into chunks amenable to management. The disciplinary forms of the university, some four hundred years in the mak-ing in North America, have created erasures so monumental that somehow universities are not able to adequately diagnose the massive, thick, intimate, knotted violence that potentially connects all relations. Confronted now by the unavoidable recognition of those violences, one can lose whole worlds in the mix. So as scholars we might feel the hunger for new moves, want to diagnose the historical moment, and build solidarities beyond walls. Yet this is not easy when so much of the university's environmental knowledge apparatus remains embedded in the production of the current enframing of planetary-scale crisis.

The geoengineering rush

As the environment merges with the planetary as a collective problem space, who has agency in a One Worldism in a state of total emergency? We fear that problems scaled so quickly and absolutely to the planetary can only invite responses equally large—such as empires, multinational companies, and geoengineering—which primarily serves to reproduce the same disciplines, militaries, and colonial relations that made the problem in the first place. After all, if the problem is the size of the planet, who has the capacity to manage it?

Perhaps it is not surprising that consensual action to reduce harm is increasingly ignored in favor of billionaire toxic masculinity agenda setting. It is revealing that one of the most common questions posed today about planetary environmental crisis is a multiple-choice question about billionaire salvation: Will it be Bill Gates, Elon Musk, or Jeff Bezos that patents geoengineering, colonizes Mars, or finally moves elites off the damaged planet?[37] Their privatized responses to the question of planetary emergency conflate an environmental salvation project with a new market potential. How can we not see today's race among billionaires to control green energy, global health, or outer space as anything less than a

---

37 See Robert Heinlein's *The Man Who Sold the Moon*, a story about a robber baron who becomes obsessed with colonizing the moon for both resource extraction and territorial control, spreading the rumor that diamonds exist in abundance on the moon to galvanize support for his first moon mission (Heinlein 2013). Heinlein's literary estate now funds the Heinlein Prize for Accomplishments in Commercial Space Activities (https://www.heinleinprize.com/), which was awarded to billionaires Elon Musk in 2011 and Jeff Bezos in 2016 for their respective programs to capitalize outer space. The billionaire dream of raiding outer space has accelerated in the twenty-first century and is fused with frontier ideologies that mirror and rival those from the nineteenth century.

1.7 (*above*) Elon Musk's launch of a Tesla into outer space just for fun, February 6, 2018. Photo, Tesla.

1.8 (*right*) Jeff Bezos's return to Earth from his first tourist trip into low-Earth orbit, July 20, 2021. Photo, Blue Origin.

reboot of the nineteenth-century race in North America to build the railroad, control global rubber, mine copper, make guns, monopolize drugs, or drill oil? Comparing them might feel too easy or on-the-nose; but ignoring their similarities would be analytic malpractice. Today, some say data is the new oil, others say attention is, or water. Regardless, "terra" remains the medium for value.[38] Terra is the medium needed to realize billionaire technocratic planetary dreams—not just in terms of infrastructures, depletion of rare earth minerals, and energy use, but also in terms of colonizing a collective future in the name of property relations and eliminating other worlds.

---

38   *Terra* is the Latin word for "earth" and, as Elden (2009, xxviii–xxix) demonstrates, also informs *territorium*, or the "place of something," and is associated with *terrere*, "to frighten or terrorize." William E. Connolly (1994, 24) concludes his review of the Latin origins of the term *territory* with "To occupy a territory is to receive sustenance and to exercise violence. Territory is land occupied by violence." See also Anidgar (2004, 54). In the colonial era, the ancient links between terra, territory, and terror were fused in the concept of terra nullius, the European conceit that land portrayed as empty was open to immediate dispossession and seizure—the foundational basis of settler colonialism; see Lindqvist (2007) and Morton-Robinson (2015). We emphasize the *terror* in *territory* as property today, but also the fulsomeness of land as the basis for a world of many worlds, that is, as a multitude of ontologically distinct forms of relationality that shape and inform communal life in relation to place.

We'll have more to say about terra in a moment. But just pause for a moment to consider that Gates funds nuclear power and geoengineering research while also being the largest farmland owner in the United States, a distinctive settler achievement.[39]

Bezos says the only way for capitalism to grow is for it to move off-world, positing the virtues of floating outer space colonies that would also open up the moon and asteroids for mining operations, restaging the frontier as an exercise in corporate salvation.[40] Musk, meanwhile, is racing to get to Mars and intimating that his first move will be to detonate nuclear weapons on the polar regions to superheat the atmosphere, exercising a male fantasy of being a literal world-maker.[41][42]

Do not be taken in by the rhetorics and aesthetics of the massive scale of environmental planetary enframing. It is not massive at all. Both too small and too big, it boasts of long reach but only delves in certain reservoirs for certain things; its cloak of planetary scale and deep time is a diversion from its narrow scope. Despite wanting to swallow the entire Earth and to relabel time itself, both the Anthropocene and the environment are thin configurations in relation to the thickness of being and the intensities making hostile conditions. Because the Anthropocene is ultimately

---

39 On plutocratic farmland consolidation in North America, see Shapiro (2021); and on Bill Gates's farmland strategy as a settler colonial project, see Estes (2021).

40 See Bezos's (2019) presentation of his Blue Origin space program involving both extraction projects and space colonies. Bezos is trying to realize Gerard O'Neill's vision of floating space colonies; see O'Neill (1976, 1977).

41 On Musk's plan for Mars, see Lockett (2020) and Wall (2019). James Rodger Fleming (2017, 23) calls most advocates of geoengineering WEIRD—that is "Western, Educated, Industrialized, Rich, and Democratic (WEIRD) males with superman complexes." The lifestyles of the richest 1 percent now contribute more to global heating than the poorest 66 percent of the global population; Laville (2023). For assessments of geoengineering scenarios, see Kintisch (2010); Keith (2013); and Hamilton (2016).

42 Geoengineering scenarios are a form of official science fiction. Consider Oliver Morton's (2016) conclusion that geoengineering is inevitable as superheated cities in the Global South will turn to it for relief. See also Kim Stanley Robinson's (2020) exploration of a similar scenario in *The Ministry for the Future*, or Neal Stephenson's (2021) version in *Termination Shock*, in which a gun-loving Texan billionaire initiates geoengineering on his own, distributing negative effects all over the globe. In each of these scenarios, the historical production of greenhouse gasses in the Global North gets erased by the emergency conditions generated in the Global South, creating conditions for competing geoengineering attempts. See also Von Neumann (1955) for mid-twentieth-century fears about the coming "weather wars."

an expert geological judgment about marks in the strata of the earth, it cannot articulate a political struggle against concatenated violence— and it is not intended to. No matter its final articulation within the logics of geology, naming a planetary epoch will not remove a molecule of carbon from the atmosphere; it will not house a single person, defuse a bomb, make anyone sleep more safely, return Indigenous land, or create less toxic conditions. Too many environmental concepts have no explicit investment in making more habitable worlds.

But, in the widespread enthusiasm for the environment, we might read a desire for a political knowledge-making project that would redraw our scales for understanding the historical and spatial making of a damaging present, that could thematize the world-altering effects of apparently disparate processes that connect and accumulate, that could address these concerns and see them as an ever-emerging condition.[43]

So what to do with this tangle of historical and emerging problems— how can we attend to violence in its multiple configurations, while critically building alternative modes of joined-up accountability and conditions?

That was our first stop on the Intergalactic Bummer Train—involving the hard task of critically recognizing *environment*'s histories, ongoing commitments, and current velocities—while still seeking to imagine and enable less hostile worlds. Now let's pause for some introductions.

1.4   WHO, WHERE, WHAT?

Who are we, and who do we think you are?

Before moving on, you might want to know who you are getting on a train with. Please forgive us; we should have already introduced ourselves.

Who are we? Who do we think you are? And what is this book about?

---

43   For a review of Anthropocene concepts, counterconcepts, and debates, see Haraway (2015); Bonneuil and Fressoz (2017); and Davis and Todd (2017). On the challenge the Anthropocene as concept poses to contemporary thought, see Hamilton (2016); Cooper, Brown, and Price (2018); and Coudrain, Le Duff, and Mitja (2022). See also Chakrabarty (2021) for a reassessment of the concept of the planetary for the humanities and social sciences in light of both global warming and the Anthropocene debates.

We are four professors who work at universities in the United States and Canada. We all take part in the field of science and technology studies and study questions of environment in some way. Some of us are also located in anthropology, some in history, some in geography, and some in feminist studies, and some sometimes in American studies. We have tenure. Some of us have done research in East Asia and South Asia, but we are mostly grounded in work in North America, and we all work and live there. We all went to graduate school in the United States. We are all profoundly indebted to feminist science and technology studies for our thinking. We all struggle with our disciplines and the knowledges they create and inhibit. We feel these struggles in the classes we teach. We are all masculine (but not only), and mostly White, but also Asian American and Indigenous. As tenured university professors, we all enjoy to some large degree the dividends of Whiteness, even if we may also unevenly face its hostilities. We write this book as beneficiaries of the security granted to us by North American research universities and hence from the tunnel vision created by this starting place. We hold in our bodies over one hundred and twenty years of combined disciplining inside universities; this may be why unlearning its ways is for us so effortful and strained. We are all unavoidably both inside and navigating the problems of gender in the university as we seek to undo the norms of White possession in research and cultivate modalities of joined-up difference in our modes of study. One of us is a farmer, one of us likes to draw, one is a media junkie, another has a lab, some of us are queer, and all of us think a lot about chemicals, contamination, and pollution. We are each positioned and called in our anticolonialism differently.

We did not join up to write a book—far from it, in fact. We came together because the North American corporate-captured university did not feel right. Each of us was unhappy with conventions and habits that dominate studies of environment, environmental violence, and technoscience. We also worried about the states of our worlds and how our universities worked against them. While we had no plan for an outcome, we knew that we wanted ways to meet, so we could think together even though we live far apart, and that we wanted to build something shared with each other and with students. We also knew that we think differently from each other; we know and gravitate toward different kinds of things, with often divergent sensibilities and commitments. Our hope was that feeling our way toward intellectual friendship and joined-up study might be a good way to unlearn and learn things differently. We wanted to make and share spaces that worked differently, where we might circle what feels wrong and what might be better, and stretch toward rendering those without the demands of performing mastery.

Living and working in the United States and Canada, where in the last decade White Supremacy continues to make breathtaking gains in formal political arenas as well as the currents of the everyday, and where our neighborhoods and mediascapes offered daily substantiation of the killability and forgettability of Indigenous, Black, Brown, Asian, poor, and 2SLGBTQ+ lives, we quickly found that thinking together brought feeling together with it. Intellectual friendship meant not only reading and thinking but also mourning and laughing together. Making spaces that worked differently meant making room for sadness, rage, and numbness during difficult collective and personal times. Finding ways to meet meant finding ways to hold each other or to give each other space or time. And learning to unlearn together meant not only learning to understand each other but also how to recognize and find words for what hung between us, our tensions, and writing to remember how to get there again. This book is the surprising artifact of this process of learning to think together. Its thought, affect, and situation are an irreducible compound.

What did this process look like, and who did we turn to for guidance? We started with a suspicion about the sudden and urgent, massive scaling-up of environmental problems and the use of data visualizations to constitute claims on the planetary in ways that reinforced disciplinary hierarchies in the university. We sought to think about how collective problems are always nonetheless experienced unequally, how concentrated violences have deep structural histories that were easily erased in the jump to planetary-species claims—and worried very much about the enthusiasms in the university for making precisely that jump. We organized around the theme of "engineered worlds" via a 2015 graduate seminar and later conference in Chicago to amplify this question while exploring the monumental damages of industrial toxicities and their various histories and futurities.

Thinking and studying with students has been central all along. We taught together—requesting coinciding teaching times from our universities to combine our individual classes informally into a quad-campus seminar (linked via Zoom and Piazza). This allowed us to maintain contact and continue threads of thought during the busyness of university work, but more importantly, it expanded our circle of conversation and our accountabilities to cohorts of astute students, taking up the speculative seminar themes "Alterlife, Conditions and Aftermaths" and "Terraformations."[44] Through quad-campus discussion boards about weekly readings, video calls linking our seminar rooms at the start of class sessions, and in-person follow-up discussions in our individual classrooms, we broached the project of building alternative genealogies for science and technology studies and environmental research. We were looking for concepts and questions concerning the political making of substrates for living that avowed began from material residues of and reactions to colonial violence rather than from narratives of labs or scientists or humanity. Our courses led to conferences that brought together some of the people we were reading and learning to study environment differently with. Our in-person conferencing was stopped in spring 2020, like so much, because of COVID-19, which shut down campuses and travel but not our collaboration.

Through this collective and joined-up learning, we sought to perturb our disciplinary trainings and found it meaningful to write together as well as teach, read, and talk. We met in person, in writing retreats, and online. We have each read, rewritten, and discussed every word and sentence of this book, creating something we would not have done individually, something that has unresolved differences in it, and a text that has most of all been a process of colearning across differences. The text is the result of this elaborate and generous cowriting, struggling with and made possible by Google Docs. We learned the value of spending hours carefully drafting a perfect sentence only to watch it slowly retreat from the page, radically rewritten, or be struck entirely. Marginal comments not only left breadcrumbs for future research and writing but also offered some brutal takedowns, jokes about the weirdness of our project, and of course the unruly status of life.

The result is a funny text that could only be made this way. We have worked through disagreements in long multiday and even multiyear discussions, seeking to not give in to the impulse to smooth over that which is unresolvable. We have written in fits and starts, starting and starting over this book, whenever we could get together, prompted by the insights of our shared classes and meetings with other researchers, while being batted around by and attempting to respond to changing political conditions in our universities, in our surroundings, and in our intimate lives. We have worked together through some difficult political and personal events. Along the way, we have walked, shared meals, ranted, laughed, cried, aged, and altered our minds and feelings.

---

44 Murphy (2017a) and Choy and Zee (2015) were inspirations for these courses.

Since thinking and studying with students has been crucial to this project, we imagine that we are writing this book as a way to continue that arc. We have taught each other's students across our disciplines, confusing some and always being humbled by what they teach us. Thus, we imagine that you are likely a dissatisfied researcher of the environment as well as possibly a student chafing at your discipline's colonial mode, or perhaps someone who comes to the strangeness of academia out of political organizing, and that you are looking for better ways to study massive violence that don't reproduce that violence but instead contribute to making less hostile worlds. We also imagine that you might be someone who feels very uncomfortable in the university, someone whom the university is not built for, and who may have other ways of being, learning, and organizing from your own life that the university refuses to value. We imagine that you too might feel rage about how the university operates, the predatory force of Whiteness in North American institutions, or normalized support for militarism and imperialism. Or we imagine you might be someone who cares deeply for land and repudiates violence but who feels stuck in the logics that the university wants you to participate in and are not quite sure how to do something else.

We may well be imagining you wrongly. But we wanted to situate ourselves and our mode of address as part of our resolve to study in a different way. In any case, welcome—we are glad you are here. There is always an open seat on the Intergalactic Bummer Train. Be sure to say hello to fellow travelers and share some of your jokes.

In studying with each other and struggling with changing and yet obdurate political and material conditions, we have learned from a legacy of powerful scholarship and writing that has done this better than we will ever do. As part of our process, we have attempted to study in joined-up modes from places of support, teaching, honesty, solidarity, humbleness, laughter, gratitude, noninnocence, and camaraderie. But it is also true that negative affects, such as despondency, anger, fear, fatigue, and uncertainty, have informed our work.[45] The despondency can make it hard to explicate in the manner that the university asks of us, that is,

---

45 In thinking about the challenges of study under violent conditions, we are particularly grateful for guidance in how to do affectively charged political intellectual work from the examples of bell hooks, Frantz Fanon, Fred Moten, Sandy Grande, James Baldwin, Donna Haraway, Winona LaDuke, Vine Deloria, the Combahee River Collective, Robin D. G. Kelley, and Ruth Wilson Gilmore.

clearly, calmly, analytically, as single persons located in academic disciplines that slice up the world in a way that is constitutive of the problem. We have struggled with the affects that make up our joined-up work, just as destabilizing ecological, political, financial, and health conditions informed our actions and imaginaries, and challenged us individually.

Together, in our small ways, we have also reached for a portal into something else, another university that is possible.[46] We have relied upon many models and inspirations. Together with each other and our students as a still broader collective, these models have reshaped us and some of the routes, moves, and forms that became this book. What became clear was that we were never teaching what we knew per se but were responding and becoming anew together and with the many other words and politics of others we studied. For example, reading Stefano Harney and Fred Moten, we studied techniques for carving a pocket for study without calling it to order; reading Eve Tuck, we studied when to say no to disciplinary inheritances and habits that document the extensiveness of harm and to think capaciously about what research might do when animated by desire toward potential rather than compulsion to map damages. Reading Aileen Moreton-Robinson, we studied the settler ontologies of White possession. From la paperson, we studied to recognize the traces of the future and the elsewhere already incipient in the here and now, poised to reassemble. From ongoing struggle and scholarship against environmental racism undertaken by Robert Bullard, Laura Pulido, Barbara Allen, and Devon Peña, we follow the insistence on being led by grounded political movements that understand environmental violence as the work of racial capitalism extending within the built environment of cities. Reading Frantz Fanon, we learned about being unfaithful to genres and how to work through and beyond pessimism. Reading Sylvia Wynter, we learned to tackle the order of things and its genre of inhumanity through its grammarians, the professional researchers inside and around us who study, teach, and execute its orders.[47] Be-

---

46 On what a third university might look like, see la paperson (2017); on rethinking the logics of study in the university, see Harney and Moten (2013); and on what decolonizing the university might actually entail, see Tuck and Yang (2012) and Smith, Tuck, and Yang (2018).

47 Reflecting on more than a century of dedicated antiracist efforts to remake the university, Joshua Meyers (2023, 8) identifies Black Study as "the tradition of refusal of the knowledge of the world as it was given to us by those committed to colonial

yond the works themselves, there were the conversations they sparked, run through with the lives and knowledges of students and collaborators that all added up to our joint unlearning and study toward other ways and worlds. This all may help explain how we ended up writing about study and not just against the environment but for terraformations.

This book doesn't arrive fully at other worlds, but we reach for them and emphasize them nonetheless. Just diagnosing the problem is not enough, but escape is also not an option. As a result, we have written a strange book that insistently worries about the analytic tools we use, reassessing their histories, while also offering rants, absurdisms, sledgehammers, bummer trains, propositions, slowdowns, takedowns, resets, time travel, humor, and breaking the fourth wall. In other words, we have been unfaithful to genres, and at the same time we have been in them in strange ways. Our writing together has not just been about making tools of analysis for us but also psychic and emotional resources—modes of learning together to survive and not just reproduce the problem. We are trying to be up-front about what our practices and concerns have been, what we have tried to do, and its limits. We hide neither the depressions, madness, frustration, nor failures, nor the fact that it has also been convivial along the way.

Where are we going?

And what is this book? This book is meant as a small methodological wedge into something different for us—and hopefully for you. It offers tools, terms, steps, moves, stances, and shapes from a decade-long experiment in collective thinking, teaching, and writing—elements for an always incomplete method of study that only emerged in composite,

---

and racial order—and all the ways we still experience it, the many othering practices it generated. It is a refusal of the blessing of liberal humanism and its variants, the philosophy of life and living that is really only about the political same, a violent reanimation of the status quo, the Western conceptions of what has and should always be. It is in refusing that we created Black Study as the places—in the margins and contentedly so—that challenged everything the university handed down to us as the only possible reality." Studying with and next to Black Study, we join up with the ways Meyers connects the disciplinary formations of the university to what we call "conditions" (that is, the quality of air, water, food, shelter, opportunity, and danger that inform the quality of life and relation in any given moment), making the simple but powerful case that study is an essential form of world-making.

through joined-up work, when we released the goal of individual academic mastery and worked to experimentally change our coordinates and conditions.[48] We share this way of study, which we somewhat jokingly call *terraformatics*, organizing the book into parts, each seeking to locate and articulate new methods that move toward more habitable worlds. We use *terraformations* as a central term in this book to name a provisional ensemble of material, earthly, living, social, affective, imaginative, machinic, spiritual, political, and epistemic processes. Instead of studying the environment, *terraformations* pushes us to study particular interconnected processes that create material and affective conditions for being and knowing that we are not outside of. In orienting toward *terra* and *formation*, we are committing to place-based, obligated nonuniversal study in a world of many worlds scaled at the level of the intimate as well as the massive.

Asking "What is an X?" we begin by thinking differently about two key figurations in contemporary planetary-scaled environmental science: the core and the species. By undoing them as bounded units, we hope to make clear how it is that the notions of the environment depend on histories baked into the very structure and form of near-ubiquitous concepts.

In the next part of the book, we lay out a situated premise that undergirds our study of terraformations. From our vantage point of North American political turbulence and the university, we propose that these proliferating responses to planetary emergency in the twenty-first century are elements in a provincial syndrome we call Fear of a Dead White Planet (FDWP). Liberal and illiberal forms of FDWP may appear to oppose each other (as, say, climate agreements and aggressive extraction), but they demand in common an ongoing terraforming of White possession, by which we mean an active ongoing material and social installation of particular colonial, racial capitalist, and patriarchal infrastructures into lands and lives.[49]

---

48   On joined-up thinking, see Agyeman et al. (2012), as well as Di Chiro (2008).

49   Part of the enthusiasm for planetary thinking today is that it imposes a One Worldism grounded in a singular economy and a technocratic vision of the earth as a system of systems. The establishment of a planet-species relation often erases the multitude of different ways of living with and on land that coexist, while it defines and implements the terms for a managerialism that reinforces existing corporate, military, and state structures. The earth system and the geoengineer are the emblematic forms of this One Worldism.

Next, how do we make room for something else? This is the question addressed in the next section. Writing this from within and against anthropology, history, geography, science and technology studies, and social theory, we take up these concerns in the belly of the academic beast. We look for ways to proceed with and against sanctioned methods of research. We offer some starting cues, preliminary propositions that we have found helpful for framing positive and negative spaces for research and study that are not fully caught in the spin cycles of FDWP.

What does a portal to an otherwise look like? This section of the book starts in the middle and presents repurposed objects salvaged from the proliferations of FDWP. What is land, or a lung, or a virus, or thinking? How can we put our commitments into practice, step off the disciplinary rails, and reevaluate conditions? What modes of relation surface when not rehearsing disciplinary norms and seeking actively to acknowledge the ongoing violence of FDWP?

In the following section, we land and ground ourselves in terraformatics. Here we try to spell out and theorize more precisely terraforming as another modality of study. We wrestle with the ways we dwell and act in conditions composed through mines, planning, wires, logistical supply chains, plants, nonhuman beings, potential and tapped energies, laws, winds, and media networks—material contours for worlding in a multitude of ways. The One Planet is a particular attempt to imagine and do terraforming as something fundamentally monumental, but we want to take terraforming back and make it more harm reductive, modest, and collective.[50] Through this latter sense of terraforming, we reach for better modes of study aimed at less harmful worldings.

After the heavy lifting of terraformatics, we get more practical with some methods. Heartened by good company, we invite you to join us in some modest efforts to cultivate and recognize alternate worlds, some always with us, others nearby, and conjure some not yet existing ones. Working from inside the problem with necessarily inadequate tools, we seek your company and help for the difficult but essential work of cultivating better conditions.

---

50   For example, consider the monumental achievement of the 1989 Montreal Protocol in repairing the ozone layer; see WMO (2022). This planetary process was not achieved by injecting chemicals into the atmosphere but rather by eliminating the use of harmful ones (e.g., hydrofluorocarbons); see United Nations Environment (2023).

So take a moment to breathe and relax, and then let's hop back on the Intergalactic Bummer Train to reconsider some of our objects of study. Let's start taking this train apart and making it something else. We are going to start by asking "What is an X?"—and don't worry, you already have a ticket. Feel free to bring your monkey wrench and a blowtorch.

What is an X?

Whatever discipline you are in, there is a good chance that it expects your study to begin with a known phenomenon or an object as an exemplar of that phenomenon. So what do you do when you cannot trust the objects or the methods and concepts through which they are known and studied?

Try disavowing the object you are given. We have learned to start by asking: What is an X? Where does it end and begin? What histories made it possible? From what conditions was it born? What does it enable? What interdependencies does it rely on? What does it erase? What does it destroy? Quickly the X will implode into a galaxy of relations; it will become grounded into the contexts of its making, and it will loosen into whorls of possibility.[51] Yet unfurling complexity is not the end of study itself. This is not simply because everything is complex, but more importantly, because complexity can be a beautiful booby trap. To diagram the perfect complexity map is just to remap and reproduce a world, which in the university is more often than not the One Worldism supporting fossil fuels and nuclear nationalism. To extract and explicate all the relations is to reenact extraction.

To ask "What is an X?" is not to take a bird's-eye view. It is not to step outside the problem or choose the point of entry/analytic/scale. It is not possible to disentangle from that which you study. This nonnegotiability of relation informs and torques terraforming study, not just as an analytic but as a set of responsibilities.

In other words, the relations making up X are not abstract. They are not objects. They are materially extensive, political, and ethical. They are responsibilities and flesh. Note, when we say that relations are the

---

51 See Dumit (2014) for an indispensable guide to writing an implosion experiment; and Dumit (2021) for how to understand the world in and through bromine. See also Papadopoulos et al. (2021) for a comparative study of rethinking worlds through elements.

obligations, ethics, and knots of terraforming, we do not mean that ter-raforming relations are simply (inter)personal. Instead, we are attempt-ing to approach relations—including ours with X and those whom X touches—with the sensibility that they are themselves infrastructural and connecting, political and processual. So the goal is not to map rela-tions, designate units, or add layers. The goal is to build better relations and conditions. The goal is worlding. Try shifting one degree, taking one step, into better relations with others in conditions. As conditions shift, stumble again. Let's now examine two infrastructural concepts of the current species-planet universalism.

1.5 WHAT IS A CORE / WHAT ARE WORLDS?

69°03'N, 49°54'W. A lake in Kalaallit Nunaat, also known as Greenland, recently reborn from a newly retreating glacier where the tenacious mud at the top of the ice core became captured and held the riotous jetsam and flotsam of late liberal modernity.[52] Among the sticky muck itself ground from massive forces: sweet, gentle breezes, tenacious winds, per-sistent tectonic forces halting and grinding, and the flows of rivers and tides, water and sands. Sedimented remains, returning and reanimating the textures of their origins.

What is a core? For the geologists, the core is a sample of layers of earth; it is a time-making machine and a vital aspect of constituting the earth system as a knowable entity. If the core sample works, it gives a pro-cession of linear time cut into a stack of discrete periods, telling a story about the planet—linking terra, air, ice, and water. This narrow tubular slice of soil on top of ice becomes a greedy grab, turning this pile of rub-ble and remains into a scientific planetary narrative, forcing fragments pressed together in intimate tumbles into a deep time and One World scale. Increasingly, the geological past as planetary history is mobilized via mud and ice, feeding big data sets that project distant climactic fu-tures via complex simulations of the earth system[53] Particular particles and collectivities of crumbs that fall together in just the right order at a

---

52 For a technical assessment of the anthropogenic markers in this ice core from Greenland, see Waters et al. (2016).

53 See, for example, Marcott and Shakun (2021) for a detailed study of paleoclimate as a way to predict climate futures.

1.9 Evaluation of lake core sediment from Greenland by Anthropocene investigators. From Waters et al. (2016).

remote lake are turned into a geologist's dream of deep time, universal coherence, and disciplinary dominance.[54] Out of the mud a new planetary epoch can be proclaimed, and geology gets to do the naming. This gooey mud core, centrifuged and made elemental, makes measurable plastic polymers and plutonium, carbon and pesticides, reactive nitro-

---

54 On Crawford Lake as a primary site for the Anthropocene designation, see Kaplan et al. (2023). See also Alley (2014) on ice cores as time machines enabling visions of both deep time and climate futures.

gen and pulverized fly ash into the markers of a singular planetary story, rather than the uneven and brutal unfoldings of a long run of colonial expansion, capitalist exuberance, or militarist might. The bits in the core have been marshaled into a new Charismatic Mega Concept (CMC)—the Anthropocene—that renames the condition of the entire Earth as post-industrial.[55] Through it, we are all now in a single timescape of discard, one blamed on the human, another singularity. This CMC wraps the planet, it pulls everything in and smothers everyone, and thus all and none are accountable. The human-planetary relation is compelling precisely as it erases the means of its production, the science, the consumption, the racist dehumanization, the specific forms of living that unevenly constitute and amplify injury across generations.

It is important to recall here the epistemological ground we traversed above: that our current understanding of the environment derives from the One Worldist study of military/industrial damage to land, air, and water that seeks to gather into one picture the interconnectedness of capital, empire, lichens, and being. The core is a way to sample and measure the earth system, which terminologically nominates a single interrelation of spheres of earthliness, atmosphere, and feedback loops that systematically combine to create the livable space on the planet. Within the sciences, spheres now seem to roll out like well-mapped territories, jigsaw pieces of planetary form—biosphere, cryosphere, geosphere, atmosphere, hydrosphere, lithosphere, pedosphere—that align with technical precision and a claim on totality. The earth system is made possible as a concern because it is also a digital mediation, a product of a vast set of data collection systems computational models, and crafted virtualizations. It is a conceptual achievement rendered through a layering of historical technical infrastructures (like fiber optic cables set over former train tracks), technologies that derive from militarism, capitalist logistics, and colonial occupation. The earth system then subsumes histories of settler colonial dispossession, anti-Black racism, and immigrant forced labor into the laundered futurities of big data in the digital cloud, ren-

---

55 Devra Kleiman and John Seidensticker (1985) coined the term "charismatic megafauna" to highlight the political power of pandas as animals good for political mobilizations because of their inherent charm. See Reddy (2014) for an insightful early designation of the Anthropocene as a "charismatic mega category" for similar reasons.

dering those violences impossible to see.[56] The aforementioned vaunted golden spike of the transcontinental railroad reimagined today as the geological index of the Anthropocene CMC exemplifies this, reenacting the violent connecting of labor, toxicity, and dispossession in recent blistering acts of world-breaking as a neutral method for achieving contemporary One World scientific naming.

The earth system and its approach to the mud, moreover, is a kind of nuclear fallout.[57] The twentieth-century arms race between nuclear states powers the production of supercomputers and seismic monitoring stations, where efforts to discern nuclear detonations from other seismic activity enable a basic understanding of continental drift. The atmospheric systems for measuring radioactive fallout become a network of weather stations, enabling a long-term view, transforming weather into climate. And satellites launched to spy on military rivals and to establish command and control of the increasingly digital nuclear battlefield become part of an elaborate system of planetary visualization, offering remote viewing of carbon, ice cap melt, storm surges, and smog. The frenzy of ice and mud core sampling in Greenland is enabled by a repurposed Cold War military Distant Early Warning Line station originally built to detect incoming Soviet bombers. The earth system concept begins as an American-centric vision of planetary-scale surveillance, a literal effort to track the radioactive fallout of nuclear detonations. Put another way, an anxious fantasy of following the half-lives of designed military destruction stages, funds, and girds the multidisciplinary investigation of atmosphere, ocean, geology, and biology.[58] Technically and affectively, the patterned injuries of uranium extraction, radioactive colonization, and hemispheric exposure that enable nuclear war condition the establishment of vast technical infrastructures that create the environment,

----

56　On composition of technical infrastructures and their overlapping historical forms, see Hu (2016); Starosielski (2015); and Edwards (2010). On digital media platforms, see Wickbert and Gardebo (2022).

57　For the formative impact of Cold War sciences, see Oreskes and Le Grand (2003) on seismic monitoring; Masco (2014, 2021b) and Edwards (2010) on how weather becomes climate via nuclear fallout monitoring; Turchetti and Roberts (2014) on satellite monitoring of bombs and ecologies; and Farish (2010) on the co-constituting geostrategic nature of militarism, environment, and empire.

58　On the imbrication of earth sciences and nuclear nationalism, see Doel (2003); Hamblin (2011, 2013); Martin (2018); Oreskes (2021); and Masco (2014, 2021b).

fomenting expert desires to scale ever upward, assembling an imaginary of the planetary itself.

But the earth system is more than a computational relation for knowing planetary Earth as such; it is also in and of the earth, literally. Trace the mineral veins that sustain the digital cloud's capacitors, and the earth system as computational object comes into focus as itself an earthly and earth-shaping system, with spectacular feedback loops of extraction, supply chain logistics, manufacturing, digital platforms, and computational visualizations, as well as the dumping of outdated tech. The earth system is thus both an artifact of and an engine for geological transformation—simultaneously a zone of expert desire, military competition, and phantasmic projections of precise resolution and control built out of contemporary extractive relations. The quest to apprehend earthly process as a set of overlapping and nested scales—molecular, regional, hemispheric, and planetary—is rendered possible via the digital processing of big data but also informed by long-standing desires for perfect optics, surveillance, and mapping. Big data is a product of the violence of the petrochemical, nuclear state, but data's mud and blood are cleaned by algorithmic processing, transforming decades of damaged relations into today's portraits of a breathing planet, a warming planet, a flooding planet, a migrating planet. So how should one approach the mud?

What happens if we try to remember and account for the extraction regimes, the militarized ambitions, and the corporate violence that enable the digital cloud, and Earth, to be a system? Put differently, how much of today's technology—deployed to distinguish ice, air, water, and land—are the damning technofossils of tomorrow, offering a constantly renewed opportunity to forget, to normalize, to generate the violent externality?[59] And how can we not forget this? The mapping of Earth as system, the focus on the recursivity of domains and chemical processes, the core as the sample, tries to ride on top of the sharp, slow, and enduring violences of capitalism baked into ways of knowing, even as scientists seek to document the damage. To do something different requires attending to the material histories and disallowing the externalities or erasures of epistemic and infrastructural relations that make possible earth system

---

59   For the geological sciences point of view on technofossils, see Zalasiewicz et al. (2014); and for a broader theorization of technofossils within media ecology, see Parikka (2015).

study. Instead we could mobilize particular place-based alter-histories to hold open the promise of different social relations and mobilize different futurities. To hold on to the muddy nonknown or the formally excluded, to listen for the perturbations while responding to the map of the earth system, is, however, a troubling challenge, though the mud has many qualities, stories, and mysteries that help.

For example, let's look again at the core sample. To count toward the Anthropocene for geologists, evidence in the core must adhere to a particular disciplinary form: It must be planetary in scope; it must be layered in stratigraphy; it must be a signal visible even to a visitor from outer space, that is, not an interpretation but a hard material fact—literally something you could "hit with a hammer."[60] Microplastics, nuclear isotopes, and carbon dioxide are held together in mud and ice, becoming signals of a common planetary condition within the core sample. Microplastics, fragments broken down in the tumble of waves and winds, are taken together as a mass so that they might, when aggregated, count as a single stratigraphic indicator about the planet. And it is truly astonishing to learn the numbers: "The cumulative amount produced as of 2015 is of the order of 5 billion tons, which is enough to wrap the Earth in a layer of cling film, or plastic wrap."[61] By 2050 it is forecast that the plastic cling film could envelop the Earth six times over. If the numbers boggle, the stratigraphic indicator orders. In the world of ice-as-core, the ice is a medium preserving a planetary plastic signal in time. But as Métis scholar Zoe Todd reminds us, plastic is congealed time, made from the remains of ancient beings; these remains were safely located underground until extracted and weaponized as oil. Oil and the plastic derived from it are ancestral forms, made from past living beings, not inert substances without kinships.[62]

If we begin in the crumbs of the core sample in the Charismatic Mega Concept, our research will end in the dreamscape problem of an even distribution of debris encircling a whole, singular planet. The earth

---

60   See Zalasiewicz et al. (2019, 3) on this point: a golden spike for geology is a physical location which limits the scope of the Anthropocene designation to measurable properties rather than processes.

61   See Zalasiewicz et al. (2016) for a discussion of plastic as a potential Anthropocene signal.

62   For an alternate theorization of what the fossil aspect of petrochemical forms might ethically entail, see Todd (2017a).

system will have been accomplished and a future terra summoned yet again. Microplastic is matter. Yes. It was made at a moment in time. Yes. And? And it is caught in looping reincarnations of violence, including the harms of epistemic habits. The core sample as such supports concepts that are yoked to making both earth and Earth manageable and profitable. So let's start again somewhere else, in an elsewhere that is also here in this plug of mud.

So let's start again. *What are microplastics?* Why do we think we know what microplastics are, where they begin and end, how they distribute harms and profits, what potentials and futures they might attach to, what extensive and looping timescapes they participate in, how they feel collecting in a gut? The microplastic, given a chance, disobeys the strata of a singular epoch and stretches back to the deep time of fossil fuel, millions of years pressed into new activity, pumped through risky chance-scapes of pipelines, into the refinery and thus out into the emissions, and hence into the lungs of neighboring beings, and beyond into atmospheric plumes. The microplastic is present in the commodity and its arrangements of profits and harms, dead labor and lively abundance; it is present in the structures of disposability, that which is made only to be tossed away, diminishing both objects and life chances. Microplastics are there in the corporate bottom line, the decision to make the packaging thinner, more brittle, cheaper. It is there as a companion to currents and eddies of water bodies, bonding within clays in the horizons and lines of oceans and bogs, traveling in the smog, becoming one within others.[63]

Microplastic is a strange attractor: persistent organic pollutants like to hitch rides on microplastics, finding new ways to concentrate, move into guts, pass through membranes, bind with receptors, alter metabolisms and genders, arrange susceptibilities to injury, distribute exposure to death. The microplastic breaks up into easily blown particles and yet threatens to endure past the lifespan of the fishes, the humans, the glacier, the lake. The term *microplastic* is its own bundle, holding together many different materials and itineraries as if they are a singularity rather than a complex scattering tied to a multitude of compositions, commodities, and corporations.[64] Michif scholar Max Liboiron teaches us that how we

---

63   On kin relations with water beings of all kinds, see Todd (2017a, 2017b).

64   Research has demonstrated that microplastics are now ubiquitous across earthly domains (land, sea, air) and life forms; see Rochman et al. (2019).

learn to see microplastics is also about how we learn to encounter worlds. We can learn to see that microplastic embodies colonialism because its spread is tied to a logic that designates an acceptable threshold of killing measured as the capacities of the land and bodies to absorb violence.[65] French essayist Roland Barthes noted in 1957 that plastic was the only medium that strived for mundaneness, to become the throwaway convenience of a society embracing mass consumption, planned obsolescence, and disposable culture.[66] He saw the plastisphere emerging long before the term existed, long before recycling sought to solve the problem and slowly unwrap the planet of its plastic cocoon. But the dream of a plastic or social film equally everywhere falls apart. It implodes and scales outward all at once. It is interrupted. The plastics of the laptop this is written on, of the credit card this laptop was paid for with, of the prescription glasses we stare through while reading this, are noninnocently caught in what microplastic is. The microplastic is in the gun and the grenade, it is in the cheap shoes that wash onto the shores of beaches, evidence of a regularized traffic of seaborne refugees gambling on life chances. It is the remains of ancient living beings repurposed. Starting thus, we might glean from microplastics more than a thin layer in a planetary portrait; we might instead become accountable to transforming the categories, the logics, the institutions, and practices that render individual and global injury as the externality, not the product. Is it possible to dream other worlds with microplastic? Can it recognize the other worlds already long here? It is matter that matters, conditioning conditions now on vast scales in ever-deeper time sets.

The indicator debris accumulating in the mud of Kalaallit Nunaat, around our feet, and in our lungs thus propels us backward into messy genealogies, and out into contested futures. This condition requires meth-

---

65   Max Liboiron (2021) shows that environmental protection laws are constituted to enable pollution via debates about thresholds and absorption rates and offers a decolonial science alternative based not on property but on land relations. See also Choy (2021) for a critique of the logic of "externalities" in economic reasoning; and Murphy (2021) on reimagining chemical relations.

66   On plastics as commodity form, see Barthes (2013); see also Liboiron's (2021) approach to plastic from an Indigenous science perspective; and Davis (2022) for a reading of plastics' endocrine disruption consequences. See also Jain (2013) on the embodied registers of industrial modernity in US cancer culture; and Benjamin (2016) on remaking sciences via speculation.

ods of refusal. Refusing the future that the Anthropocene CMC portends, but also refusing its pasts too. How could one disrupt the conditions of planetary-scale distributions of plutonium and plastic, linking the university to efforts to make less hostile worlds? Instead of accepting a plutonium Anthropocene or a plastic-wrapped planet, who could one join up with to shape healthier future conditions in the plural?[67] In planetary climate crisis discourse, we hear about Greenland because its great glaciers will melt and flood the oceans elsewhere. The ice core is detached from the grounded history of Kalaallit Nunaat colonization by Denmark or the successful achievement of decolonization and self-governance in 2009 of this overwhelmingly Inuit land.[68] It is detached from the particularity of glaciers and ice, their specific animacies for themselves as beings as well as placed worlding relations.[69] What is ice and mud in this anticolonial, many-worlds horizon? With whom, and with what entities, could environmentally concerned academics join up?

Accountability to this core means refusing to reify the land of Kalaallit territory being captured as merely a medium for planet-scaled ambitions, the means of making the quotidian universal, of defining the life and violence out of objects. The ice core is used to tell an unplaced deep history of a planetary ideal, scaled through military atomic testing with not a mention of local emplaced colonialism that makes the science possible in the first place, including the US military Thule Air Base on Greenland, where in 1968 a bomber carrying a nuclear payload crashed, resulting in radioactive contamination of the area.[70] What would the core become if it joined up with anticolonial futures of Inuit people of that land? Or refused military eyes? Or respected the specificity of ice and glacier as particular forms of living being independent of human domaining? What new responsibilities would emerge rather than collecting documents of injury? What does this world-of-many-worlds core teach? That mud can be made to speak in many different ways. It is the

---

67 For a discussion of plutonium and plastic as proposed Anthropocene signals, see Masco (2021a); and see United Nations (2017) on a global treaty effort to ban nuclear weapons and war.

68 On Inuit self-governance and decolonial land politics, see Lyall (2009).

69 On the animacies of glaciers and ice, see Smith (2025); Dodds and Smith (2023); and Hobart (2022). On the animacies of water see Ballestero (2019) and Todd (2017b).

70 For ecological damage assessments of the Thule nuclear incident, see Eriksson et al. (2008) and Mitchell et al. (1997).

ground out of which concepts such as the Anthropocene or the environment are wrestled, but it is so much more. It is its manifold self. Mud can be both liquid and solid, tenaciously viscus and unreservedly fluid. It holds world stories and ancestors, and confounds friction, at once deep time and quicksand, both clay pot and silted slurry. In fact, it is one of terra's many, many names; it is within many shifting relations, and no one knows all of them.

1.6   WHAT IS A SPECIES / WHAT IS A LOSS?

In 1987 in the cloud forests of Kauai on the Alakaʻi Plateau, the last air was forcefully and eloquently pushed through the lungs of a small bird in the honeyeater family known as ōʻōʻāʻā or Kauaʻi ʻōʻō (or *Moho braccatus*). It may not have been its last breath—that would come soon enough—but it was the last recording of its species, the last auditory evidence of centuries of an embodied form of sonic evolution geared to finding and connecting to others.[71] It sang solo into the void, one half of an unanswered duet, the end of a collective music of kin and kind.

This last desperate song—sung by the Kauaʻi ʻōʻō for a partner who had already perished—is what scientists call a living dead species. Its relations in community through pollination, seed dispersal, its consumptions of insects and snails were transformed by its absence. Before its haunting and rhythmic sounds and complex relations were extinguished from Hawaii's forests, they were etched through oxide particles into the magnetic plastic on a Cornell scientist's recording device. Its vocal vibrations were then transformed again, this time materialized through lithium, a rare earth element from the Salar deserts of Bolivia (where the last remaining Andean flamingos, or *Phoenicoparrus andinus*, are themselves being pushed to the brink by habitat loss), then brought back as light and transferred through fiber optic cables, now part of the muddy sediment of oceans, to be installed in the digital natural history files of Cornell University and eventually to the bowels of a YouTube video and into the clouds of the internet.

These digital clouds are material and grounded, not ethereal. In fact the Kauaʻi ʻōʻō has migrated to the wild and unregulated corporate zone

---

71   The American Bird Conservancy (2015) has posted the recording on YouTube.

1.10 Kaua'i 'ō'ō (*Moho braccatus*). Illustration by John Gerrad Keulemans, 1893.

of North Carolina where Apple can lucratively benefit from the lack of regulation on Duke City Coal to access cheap energy to cool and store its haunting calls. The Kauaʻi ʻōʻō has also been sighted on water-cooled industrial data farms in the parched deserts outside Reno, Nevada, and even in Facebook's undersea data storage facilities that avail themselves of the cool but warming oceans to cheapen its cost and fill its pockets. From these oceans, deserts, and farms, the bird's digital cloud presence is piped along five hundred miles of fiber optic cables (from silica mined in Washington state, which requires massive amounts of water and produces tons of carcinogenic particles) to San Francisco and beyond, giving 50 million people access to its song within 14 milliseconds. Far from isolated, the Kauaʻi ʻōʻō is enmeshed in multiple ecologies, economies, and socialities in both its breathing and nonbreathing forms.

Extinction, from the Latin *extinctum* ("extinguished"), from the verb *exstinguere*: no longer alight, a family having no living members, the wiping out of debt, having no valid claimant, to kill or destroy.

How is it that the loss of the Kauaʻi ʻōʻō could be rendered into a form that so flagrantly avoids the politics of its disappearance and the implications of its loss? What foundations does the concept of species extinction build on? What forms of response are compelled by its incitement to action? It's not that loss should not compel and incite, it is that species extinction seems to produce responses and relations inadequate to the moment and its manifold effects.[72]

The accounting for what is lost and what is made as the Kauaʻi ʻōʻō sings its last song is curtailed by the very concept of *species*. In relying upon species as the relevant parameter for conceiving of endangered living, it preoccupies itself with life's extinguishment only when it can be rendered at the level of the taxon.

It is no secret that the present convention of species, one of ordering life forms into nested categories, originated in European efforts to tame the chaos of colonial encounters with unknown natures. Victorian empirical observation of the sexual organs of plants sought to posit essential distinguishing qualities of collectivities: species, a concept that paired an idea of essential characteristic difference to the figure of a re-

---

72 For detailed studies of anticipated loss and extinction, see Parreñas (2018) and Van Dooren (2016, 2019); and on endangerment as a mode of future anterior projection, see Choy (2011).

Taxonomic Ranking System

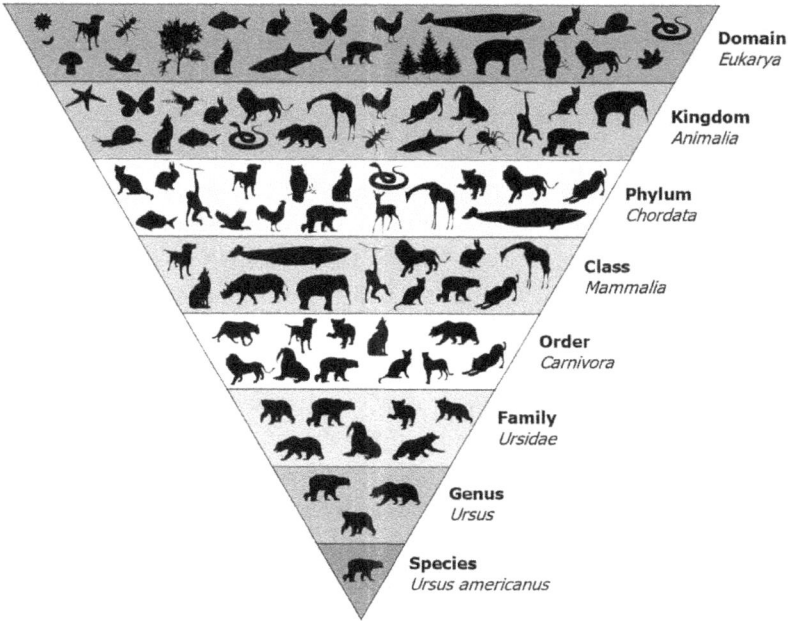

1.11 Taxonomic chart of species. From Fauna Facts, 2024.

productively isolated population. Taxonomy's technique of nesting no-menclatures, infinitely expandable, made the possibility of a situated European knowledge aspiring to universalism.[73] Colonial and commer-cial logics—of claiming through naming, of turning relations into objects and property—structure the components and methods of classification that made the species as a concept and made taxa of plants and animals available to knowledge, desire, extraction, and trade.

Histories of species extinction have accelerated at stunning rates since the invention of the species concept. Recent studies suggest that the current extinction rate is over one thousand times faster than the

---

73   For an assessment of universal knowledge making, taxonomy, and empire, see Pratt (1992). See Kosek (2010 and forthcoming) for a critical natural history of the honey bee.

average rate in Earth's history, which gives the current epoch the fastest rate of extinction on record, even if it is not yet a mass die-off. It is not the first concentration of mass species death in Earth's history—it is currently labeled as the sixth—but unlike those previous, this one is not animated by extraplanetary impacts, volcanic eruptions, or mass pandemics. Instead, we are told, the deaths all derive from the industrial aftermaths of a single species. This is the story of anthropogenic mass extinction: In the space of a few hundred years, humans have directly driven over one thousand known species extinct, and another million plant and animal species are threatened or endangered in the coming decades.[74]

So, in the conservationist's apocalypse, the species concept permeates. It structures both its figuration of lost life and its characterization of a single species collective responsible for the decline.

While the notion of anthropogenic mass extinction introduces an idiom of accountability into assessments of species loss, it simultaneously creates a smoke screen. No need to think about colonialism, imperialism, corporate profits, differences in per-capita consumption rates—mass extinction becomes a collective species responsibility. We will not rehearse the erasures and displacements effected through the scaling of accountability at a global species level; they have by now been thoroughly critiqued. Instead, we ask, why does it persist? What work does the species concept do that makes it endure and proliferate in the North American university's planetary turn even as it has been so profoundly questioned?

Perhaps the concept endures because it supports the very terms for conceiving of political decision and action. For instance, a differentiation between capacities of humans as a species from those of other species grounds questions of intention and consciousness. Through the species distinction come possibilities and responsibilities of the human as a sovereign or savior, elevated into authority as the manager and conservator of all others. The call upon humans to save other species reproduces the authority and position of the humane response, above and outside of nature and all other species, the position of the exceptional human type. This further makes humans responsible as a collective whole, a troublingly undifferentiated and unaccountable mass. It also delineates

---

74   See the calculation by the United Nations of accelerating species extinction rates in United Nations Report (2019). See also De Vos et al. (2014); and for a broad overview, Kolbert (2014).

humanity from the webs and interspecies muddles that might make possible different connections and relations and productive troubles.[75]

But what dies when a species dies? Along with the last songs of lonely island birds, we can think of the stubborn remains of two colonial elephants from India and Africa who, in the late eighteenth century, were cut into pieces and then resurrected in Paris and compared to elephant fossils by Georges Cuvier, the French naturalist. Cuvier and his elephant bones severely disrupted the earlier European belief in the stability of species, as he claimed that within the teeth and bone marrow of the slaughtered and fossilized elephants could be found distinctive structures that proved their uniqueness from each other, and he asked a simple but profound question: "What has become of these two enormous animals of which one no longer finds any living traces?"[76] How could, or why would, a benevolent and perfect God place species on the Earth only to have them disappear? With that, the bones became the living absence of a species, materializing the notion that a species form could cease to exist.

What would it mean to mourn loss without the categories and distinctions of species? Or to do so, in a way that did not reanimate the human distinction or the collective categories of species that render invisible far more diverse and interconnected forms and beings? To ask this is not to diminish a sense of loss of songs, colors, and flesh manifested through millennia of relations. On the contrary, it is to point to the inability of species extinction narratives to account for the totality of those losses and their extensive relations with the concept of species death.

We hold that the loss is much larger and much more serious than the species narrative allows. We hold that frames of species extinction are wholly inadequate to understanding the depth of interbeing interdependent loss, the breadth of the refracting silences. What complexities and relational densities might be reimagined beyond species? What political possibilities and collectivities might open through this critique of the universalized future of the sixth extinction or anthropogenic defaunation? What would a politics of loss be that surfaced enmeshment, interdependencies, and obligated vulnerability rather than delineating species and humanist savior fantasies?

---

75 For important alternative formulations of responsibility for violent conditions, see Haraway (2016b); Tsing (2015); and Tsing et al. (2017).

76 On Cuvier and elephant bones, see Rudwick (1997, 22).

What are the constellations, beyond species, that might honor other collectivities, the porousness of forms, the entanglement of relations, that species cannot hold? And how better to account for the immensity of the loss and violence without using conventional terms enrolled in ongoing colonial violence? Or how better to hold to account the conditions, institutions, and individuals that profit from these violences, frames, and silences?

So far on the Intergalactic Bummer Train we've perturbed the beloved concepts of environment and species and taken some wicked turns away from canonical modes of thinking in the university. Whew! Now that these starting points are perturbed, new beings and relations become part of the problem space; they enter and begin to pull apart the train. If you feel like pulling the emergency brake and jumping off—hang in there. Our next stop brings us to a diagnosis of the problem we are facing in the North American university and some steps toward a new mode of study.

PART 2
WHO'S AFRAID OF A DEAD WHITE PLANET?

2.1 SITUATED PREMISE—FEAR OF A DEAD WHITE PLANET

The disaster, as most people know in their bones, is not in the future; it is already here. It has been unfolding for generations, concentrated in bodies and lands, organized by systems of race, coloniality, gender, and value, built by fences and spreadsheets. So how do we explain the current sudden eruption in the North American university of apocalyptic narratives, predictions of total end times, anticipations of a final planetary death? Or to ask this more directly, who was able to live so unconcerned until now about hyperviolent earthly conditions? What modes of insulation rendered a ferocious White supremacist capitalist order as someone else's problem, a thing that, for some, could be ignored, externalized, forgotten, walled off, observed, or simply profited from? And what work is done through this novel planetary currency of concern? Experts at podiums now announce, with exhausting repetition, that the planet itself is in danger, that the human species, each and every person on the planet, must hold their hefty share of responsibility for havoc wrought across ecosystems. The refrain "we are all in it together" deflects scrutiny from the concentrated sources of installed violence in particular structures, places, companies, and biographies.

But what if we resist being rushed to the planetary, say no to the universal pull of a total crisis, and slow things down? What if there are already good responses to the problem of massively scaled violence being worked out by the communities most affected across generations? Let's press pause for a moment on the thrust to planetary ecological end times even as we feel the materiality of harm. Let's start instead with an obvious but often avoided point of departure: North America is the site of a multilayered and still-active project of violence. Founded in settler colonialism, anti-Blackness, and immigrant exclusion, it was amplified by petrocapital in an ongoing commitment to extraction and accumulation, materialized in over a century of military empire, carceral expansion, racialized labor, and supercharged by nuclear nationalism. While we were writing this book, the university has been shaped by Trumpism, racist policing, new geoengineering departments, anti-Asian violence, pipeline and mega-mining project approvals, laws criminalizing transgender lives, laws forbidding critical race studies, censorship of words like *climate change* or *gay* or *slavery* in curricula, attacks on the right to resist war in the university, the overthrowing of reproductive rights and affirmative action, genocidal wars, and so much more. North American universities—whose disciplines and campuses developed in service and relation to these formations—are caught up in and attached to a particular politics of White Supremacy and its making of planetarity. This observation is not new.[1] Nor is it a discovery. It is a basic starting point learned from a tremendous archive of resistance documenting and responding to the hostile environments of colonialism, racism, and capitalism, some of which has been produced by resistant pockets of work inside the university itself. It is the situated premise for understanding our own particular and placed conditions of study here and now. It is why we boarded the Intergalactic Bummer Train.

---

1 For a few of the key texts on North American racial formations we have learned from, see Asaka (2017); Bhandar (2018); Karuka (2019); King (2019); Lowe (2015); Ngai (2014); Omi and Winant (2015); Goldstein (2014); Kaplan (2002); Slotkin (2000); Du Bois ([1935] 1998); Drinnon (1997); Davis and Todd (2017); and Di Chiro (2017).

The environment cannot be separated from this multilayered formation. Starting from the recognition of this historically specific matrix of domination,[2] let's try to rediagnose the problems denoted as Anthropocene, planetary crisis, species extinction, and even climate change as problems of hostile conditions produced materially and epistemically by patriarchal White possession.[3] When we write *White* here, in the phrase *patriarchal White possession*, we are pointing to a historically specific project emanating from North America and Europe to reproduce a particular historical and territorial formation, not a phenotypical attribute of bodies. What is Whiteness, if it is not an attribute of bodies? It involves approaching land, beings, and people as property in order to both dispossess and extract, creating and accruing generational wealth violently protected via possessive investments in law, policing, and everyday practices.[4] Think Whiteness as a distributed formation with shareholder entitlements, one where you don't get to choose how much stock you hold even as it unevenly distributes exposures to death and generates wealth.[5] Whiteness is not abstract here. But like Planet Nine, it also moves at vast scales, exerts gravitational forces, and has many perturbations and rogue relations. Whiteness produces second- and third-order effects. In this book, where we are concerned with the North American university, we are attending to it as a particular changing and enduring structural

2   See Collins (2002) for an important theorization of the imbrication of race and gender as a matrix of domination, as well as hooks (2007). See Kosek (2006) on race and nature; and Blanchette (2020) on race, labor, and animality.

3   Consider that Republican pollster Frank Luntz advised George W. Bush to always use *climate change* over *global warming* to mute the apocalyptic implications of an overheating planet; see Lee (2003) on this point. Bush worked furiously to downplay and limit climate science, while supporting big oil by deregulating fracking and pursuing imperial resource wars under the banner of counterterror. The first Trump administration intensified these efforts by attempting to break the Environmental Protection Agency. It did so by relocating EPA offices arbitrarily to frustrate scientists, by removing the term *climate change* from federal websites, and even by trying to decommission environmental surveillance satellites to thwart data collection.

4   A "possessive investment" in Whiteness is George Lipsitz's (2006) term for US infrastructures across law, housing, policing, education, and investment that enable and protect racial hierarchy. See also Pistor (2019).

5   See Karuka (2019) on "shareholder whiteness," which links property, infrastructures, and stock markets as vehicles for promoting White advantage across generations. See also Moreton-Robinson (2015) and Harris (1993) for important explorations of racial capitalism.

form of dominion and hierarchy that manifests in the quotidian postures and desires of subjects; in the university this manifests as managers, engineers, experts, prophets, professors, administrators, designers, donors, heroes, and so on. Whiteness creates infrastructures of knowing to support an objectifying claim on land and bodies, one built out of Western rationality that aspires to a universality that erases so many other relations and ways of being. There are many forms and manifestations of Whiteness emanating from different locations and struggles, not just North America, yet its instantiations, while wildly disparate, are often mutually supporting and even sanctioned from common sources and histories.

Think of Whiteness as a breaking of some relations between lands and peoples and a making of others, the latter foundationally connected to declarations of terra nullius, to dispossession.[6] When Captain Cook arrives in Australia, he and his colonial crew meet Indigenous residents with a complex language of, and attachment to, land. Nonetheless, he declares the place empty and quickly starts renaming rivers and mountains to erase the very Indigenous presence the crew relied on for life support. Erasure was a start to claiming ownership.[7] This is terra nullius in action—constituting an impressively destructive con of Whiteness, one connected to the thingification of being, and the commodification of land, that provides an alibi to endlessly extract profit and pleasure from acts of taking and destroying. Think of this breaking and making as an ongoing material shaping that bootstraps itself through the racializing transits that transmute some people into animal or alien labor, or into objects open to harvestable injury and destruction since 1492.

This book responds to and reflects upon efforts to characterize planetary emergency—problems typically characterized as environmental—as caught up with settler colonial Whiteness. We offer two main arguments. The first is conceptual. We propose that rather than looking at environmental problems, we understand them as terraformations. Patriarchal White possession makes particularly violent terraformations. We are interested in making alternative terraformations that work against Whiteness. We will say more about this, but for now, understand terra-

---

6   See Lindqvist (2007) for a detailed account of terra nullius; and Moreton-Robinson (2015) for how it supports and informs Whiteness.
7   On Captain Cook's approach to land and terra nullius, see Carter (1989); for a theorization of colonial orders and profit-taking, see Césaire (2000).

forming as the processual and accumulative installation and destruction of more or less habitable and more or less hostile worldings.[8]

The second premise is that one of the main terraforming projects responsible for what are currently understood as environmental problems in the North American university is White Supremacy. Whiteness is an ongoing terraforming project. Crucially, in the North American university, this project extends through the very wares offered for relief, free-to-use notions of planet, environment, and species, the uses of which accrue the hidden cost of adopting their legitimating epistemologies, thereby reproducing the very structures that create massive violence in the first place. This is what we tried to show in the first part of this book. Whiteness at its most vicious wants to at once render the world an object and recruit shareholders to its terms and world-breaking until there is nothing left.[9] Thus, much of this book is concerned with our efforts not to reproduce this Whiteness in our research practices.

The epistemic and material condition of Whiteness is a domain of damnation. In the North American university it is also where we are forced to start wrestling with the noninnocent projects of researching environmental crises. There are other imperialisms and racisms striving for planetary form, to be sure. But the Intergalactic Bummer Train has to pass through here, the North American university, where the environment has been historically constituted and continues to structure academic thought. We offer a diagnosis of/hypothesis about those projects and the conditions of their emergence: Fear of a Dead White Planet (FDWP). This is our best effort to characterize the specific historical, technical, and affective processes in the North American university whereby interlocking infrastructures for massive death apportioning can not only be missed but reproduced and amplified by an assemblage of sciences and modes of study that supposedly seek the betterment of a shared

---

8 Terraforming is usually cast as a distant future project involving space travel and transformations of life on a distant planet, but for us it is actually a central process of living, one that informs existing conditions while conditioning futurities. See Beech (2009) for an introduction to the concept of terraforming; and Haraway (2016a) for a reformulation of the term as everyday world-making.

9 For an assessment of the Whiteness of contemporary end-of-the-world narratives, see Mitchell and Chaudhury (2020); for an assessment of racialized diagnoses in psychiatry, see Metzl (2011); and for vital assessments of race within US visual culture, see Smith (2004, 2020).

world. FDWP gives us a shorthand that indexes the interconnected concatenations of violence that our baroque sentences keep describing.

This problem of Whiteness reproducing violent infrastructures through authorized research is not particular to the study of environment. Importantly, White Supremacy does not have to explicitly speak racial categories to do its work; it has created a panoply of second- and third-order grammars and normalized epistemic practices that enact its death drive.[10] Whiteness can work through these perturbations, details, influences, instruments, and weird indirect trajectories as well. Like the way it could be possible to miss an entire planet in our solar system despite intensive study, North American academic disciplines habitually do not see the vast and enabling gravitational force of Whiteness in almost all they do.

In focusing on science, this book particularly attends to these second- and third-order grammars, habits, and influences that permeate the North American university, such as ways of universalizing, objectifying, naming, and scaling. When researchers take up the question of studying the environment, they often end up in these epistemological norms and interlocking conditions. So when one takes up the creative act of studying differently, it keeps getting called into a constant relation of struggle with intolerable histories or gets called out as being unrealistic or nonscholarly, undisciplined, too biased, too emotional, imprecise, not marketable, or not strategic for promotion, when it does not adhere to academic norms.

The planet and the environment, in other words, are themselves key productions of the ongoing terraforming project of North American Whiteness, concept-technologies for serving extractables and knowables, as well as stages for its self-justification and its own innocence. At its most basic, White Supremacy wants the right to kill widely with impunity. It wants to see the world as filled with inert and killable things that can become objects and commodities. It is willing to destroy everything for its own benefit. It is made up of violent desires and property relations. Whiteness is installed in the state and its laws, in the police and its prisons, in the company and its labor, in the pipeline and its ruptures, in national security and its weapons, and in the university and its disciplines. Whiteness is intimate even as it bludgeons with vulgar force.

---

10   On the grammars of race and gender within the European enlightenment and the possibilities for a radical reinscription, see Wynter (2003).

Whiteness comes with guns and also with books.[11] Whiteness wants to save a cute animal, post a meme, hide behind a false politeness or a verbal commitment to rights, but also keep its slow and fast profits going on the side without interruption.

The disaster is already here

Whiteness comes in many flexible forms. It cannot be pinned down. It is not coherent. While its flexibility and variability are linked to its ongoing success, they are also linked to its constant failure. Its sham is always being called out; it fails to gather all it wants; its universality is easily punctured by those who are not in its epicenter.

Fearing its own death and disregard, Whiteness today deploys both crisis talk and adventure capitalist commitments to creative destruction in an effort to continually remake the future on its own terms. It tries to extend its reign via projected anxious concern for others, other species' deaths, sea level rise, and carbon dioxide levels. Facebook's original motto—move fast and break things—captures the spirit of Whiteness, revealed in geoengineering, technofixes, and carbon capture schemes, illustrating a desire for a continual nervous land grab marketed as an altruistic effort to connect the world while covertly extracting world-historical profits.[12] The end-of-the-world anxiety that characterizes Whiteness can be contrasted to the grief, fate playing, and pragmatic pessimism of those of us who experience ourselves not as above or outside violent structures, but rather inside a long arc of living, despite, and even joyfully in revenge of, having already been altered, violently perturbed, or even destroyed in our beings and worldings.[13]

Within the university, surveilling its own violence through its own eyes, Whiteness does not face the truth of its own history. It cannot and will not face it. It seeks to excitedly reproduce itself by trying to materially extract itself out of flesh, terra, and being, but without naming it as such. It seeks to reproduce itself by continuously reconstituting its

---

11   On the force of White Supremacy within institutions, see Yacovone (2022); Roediger (2017); Du Bois ([1920] 1999); Baldwin (1984); and Eze (1997). See also Marx ([1867] 2004) for a theorization of the violence inherent in the commodity form.

12   For an assessment of the ambitions of social media companies, see Taplin (2017).

13   On the critical parsing of affect about climate change in relation to colonialism and Whiteness, see Sultana (2022) and Burton (2020).

conditions—a material, epistemic, and psychic surround of infrastructures, technologies, beings, concepts, practices—that are coercively and anxiously installed as the only possible world, and the only possible fix.

Foregrounding the structure of Whiteness—and Whiteness as structure—helps us here to name and deflate a resulting managerial projection of environmental end times as a superpowered and strange anxiety that runs through the university, which we want to turn away from. The anxiety of managerial White Supremacy is that it will end, that its specific political, economic, libidinal, and planetary orders will no longer reign. It has become a truism to say that it is easier to imagine the end of the world than the end to capitalism.[14] Why is that? Conservation of this White possessive world, a world that masquerades as all worlds, is a project that can only see the end of its violent reign as the literal end of the planet. We want to try to say no to this One Worlding of a problem, and to keep insisting on what is already known, that we are surrounded by invitations to other formations and to other worlds. That knowing and being can be intergalactic. To say no is not to look away from Whiteness but to analyze and critique its modes of operation, to confront its arrogant assertions, to work against while maintaining a beyond, and even a multitude of beyonds.

We return to our earlier question: How is it that centuries of intensified death—of people, animals, fish, and plants—is suddenly a shock requiring a new planetary-scaled concept and response? For whom is this surprise possible? What modes of inattention condition the sense of emergency in conventional environmental politics? In this formation of Whiteness that orders so much research, optics are out of sync with affects, or affects are deadened to optics, creating a strange dislocation of knowing today—simultaneously urgent and anachronistic, attuned and dulled, concerned and anesthetized. The new surprise about accelerating deaths is unsurprising when we note that the epistemologies of Whiteness have always relied upon violence, its institutionalized legitimization, and its continued erasure. The epistemic conventions of Whiteness are made from tricks of disembodiment, abstraction, erasing tracks, delimiting, and reducing.

---

14 See, for example, Jameson (1998, 50), as well as Chakrabarty (2009) and Masco (2021b).

Is the Anthropocene and its planetary idioms merely Whiteness talking, in which a certain kind of culture of death comes to narrate the tragic possibility of its own dying, taking all other worlds with it? That would be way too reductive and too strong a claim all at once. Captain Cook did not get to keep discovering already fully inhabited worlds, exterminating along the way—when he came to Hawaii he was finally killed by Kanaka Maoli. Moreover, the North American university is made through Whiteness, but not only. It is an important site of resistance and filled with opportunities for creativity. Moreover, the stretch of logistics, smoke plumes, and military networks are indeed vast, and they are important to trace and understand. And to simply dismiss the materiality of Earth-scaled violence would be its own absurd erasure. We want to try to figure out how to be both with and against a sense of the massively scaled, and to think seriously about the call to planetary scale as the necessary frame.

So, what about trying to address the problem up front in a way that holds the realities of massively scaled violences and formations of White possession together? Let us create descriptions for what it is and does, not what it calls itself. One such description might be Fear of a Dead White Planet.

FDWP = F + D + W + P

Fear of a Dead White Planet is our description for this epistemic and material formation of violence. FDWP takes inspiration from Public Enemy's influential 1990 rap album *Fear of a Black Planet*. As a work of political criticism and Black insurgency, the album both diagnoses the condition of anti-Blackness in the United States and participates in the activation of a Black radical tradition.[15] Honoring this work, we seek to diagnose and

---

15   The album builds on the scholarship of Frances Cress Wesling (1974), an African American psychiatrist who developed a theory of White Supremacy. Importantly, Wesling's work articulated a theory of global White Supremacy founded on the genocide of people of color, founded on a racist preoccupation with genetic survival driven by White inferiority. Musically the album was genre mixing, inventively using sounds of city life, political speech, and media samples collaged with musical samples. We also note Michael Warner's 1993 edited book *Fear of a Queer Planet* that builds on ACT UP and emergent queer politics as an insurgent project that challenges the heteronormativity of social theory (the book, however, does not acknowledge its citation of Public Enemy in the grammar of its title). See also Austin's (2013) study of fear and anti-Blackness.

dismantle the FWDP and participate in the activation of other worldings. FDWP names the way White Supremacy is anxiously constituted through fear for its own demise, and willing to kill to keep itself alive.[16] Let us break down what we mean by it. Let us begin with its Fear.

Fear. When we use the word *fear* here, it is not to describe our own pessimism about current conditions but to describe the psychic dimension of White possession that secures itself through the sublime and paralyzing worries of environmental academic research; that is, of worry about the demise of White possession itself. One of the most salient aspects of FDWP is that it is fundamentally a counter-terraformation that pretends it is saving others while it secures itself through terror and taking. In its violences against lands and people as much as its outpourings of planetary concern and global humanism, it is a counterproject against the world of many worlds that most of us are in.[17] By this, we mean that FDWP is always in a reaction formation to the prior, ongoing, and abundant presence of other modes of being, doing, and living. This counter-formation, in its paranoid reasoning and commitment to dispossession and One Worldism, imagines defensively destroying everything rather than joining the world of many worlds. It is anxious about its own demise and manifests this in part through its way of being anxious over the planet. It is important that we confront this fear and not align with it. Our pessimisms are different. We call out Whiteness's anxious fear as underwriting self-serving projects that reinstall structures of domination in the name of saving others.

Dead. FDWP is a death formation. We mean this in three ways. Most obviously, as an always anxious counterformation, FDWP fears the prospect of its own death or accountability for its practices. Also, the world that FDWP imagines and makes is through its fast and slow killings—both immediate forms of death and structural forms (toxicity, poverty, policing) that are equally relied upon. FDWP can even delight in its killings, affectively enrolling people in its death cult. Third, FDWP is a dead world in its modes of rendering people, beings, and surrounds as inert and inanimate. Beings are made into things and objects that are thereby takable and destroyable toward its own benefit. Extraction/profit/exclusion are linked formations enabled by these specific death drives.

---

16 On the concept of countersovereignty, see Vimalassery (2014) and Karuka (2019).
17 For some comparative examples, see de la Cadena and Blaser (2018).

White. FDWP is a formation of Whiteness and White Supremacy with a particular history and set of operations. North American Whiteness as a racial project enrolls and recruits as its entitlement to do whatever it wants to that which is not itself. By Whiteness we do not mean an attribute of the subject or of bodies, but rather something that installs the individualized subject, the species, settler sovereignty, heteronormativity, property, and economy as the horizon of its colonizing One World. It is the inherited crusading positions of a singular Christian God and his truth above and outside all others, a position and power transferred to the absolute sovereignty of the king's total dominion over all subjects and territory, and then later given to the colonial charter company, and today the state. It is an inheritance that positions the specificity of Christian, enlightened ensoulment as different from and above animal, land, other. The civilized over the savage, master above slave, patriarch over family, owner over land are all an ongoing changing claim to inherited privilege. This sense of Whiteness endures through the righteous swagger of class, the naturalized reason of the market, and land reduced to resources and commodity.[18] Whiteness is the inheritor of these presumptions and positions. Whiteness is possessive in its violence, and it is nonconsensual in its operations. It wants to reign supreme, by any means necessary. Whiteness unevenly distributes its dividends and spoils, even as it recruits by offering you one measly share in its benefits. It promises that it will save you, while at the same time killing you.

Planet. To say that FDWP concerns itself with the planet is to recognize the imperial wish-fantasy of a knowable and totally controllable inert global space. The planet of FDWP is not a rock orbiting the sun. It is the concatenation of images, rhetorics, logics, and sciences of planetary systems constituting the worldview of FDWP. FDWP posits the planetary as a mode of total possession. We call out the making of a planetary formation via US militarized surveillance sciences and cybernetic fantasies of command and control situated in military machines. Planet is the One World posited by FDWP, key to its countersovereignty, reveling in desires for managerial control based in absolute exclusion of other knowledges, land relations, and modes of being. Our attention to the planetary is not against earth or interconnection, but a rejection of this particular fram-

---

18 On this type of necropolitics, see Mbembe (2017); as well as McClintock (1995); Moreton-Robinson (2015); and Bhandar (2018).

ing of the global as a managerial space and zone for Whiteness and its optics.[19] The planetary turn is endlessly swallowed and regurgitated whole by Anthropocene projects, by environmental humanities, by geoengineering, and outer-space venture capital. In naming the planetary as part of an FDWP formation, we refuse to reify the planet as the One World or as a system that can be reduced to operations, or services, or management. We also provincialize FDWP in North America: It comes from places and particular histories. It wants to claim title to the universe, but it is not actually all encompassing, even if it is massive. FDWP has an acquisitive relation to the planet, but we refuse to accept its invitation and claim to be master of the One World. We prefer the challenge of joined-up efforts of an Intergalactic Bummer Train and the insistence on many worlds, complex emplacements, and interdependencies.

In understanding FDWP, we learn from a thick itinerary of activist antiracist scholarship that resists White Supremacy as an extensive worlding and world-breaking project. Fields of study are also force fields—filled with perturbations that are working on us in both overt and subtle ways. Here are the influences we recognized—we don't know all of them. We are not all in all of them. We build on them in aspiration, influenced by these complex texts and histories in an ethics of joined-up engagement. We feel them in our thoughts and are thankful to have these forces in the university that can perturb our thinking in these ways. As early as 1940, W. E. B. Du Bois asked why anyone would desire Whiteness, concluding it promises "ownership of the world, forever and ever, Amen!"[20] Black studies has demonstrated how White Supremacy is an

---

19 Gayatri Chakravorti Spivak (2003, 2015) marks the planetary as a potentially revolutionary domain operating beyond the colonial legacies of the nation-state, international order, and existing languages but also one that is not fully knowable—a strict rejection of any managerial planetarity. For Spivak, the planetary is both untranslatable and fundamentally other to all those historically conditioned by colonial orders. Chakrabarty (2009, 2021), by contrast, offers a reassessment of the planetary drawn from earth systems science, embracing a singular planetary climate emergency.

20 In Du Bois's autobiographical *Dusk of Dawn* ([1940] 1983, 30), he precisely names the totalizing deadly acquisitive worlding of Whiteness. Whiteness is epistemically a fantastical "horizon of perception" but also a violent domaining that claims "title to the universe" and promises a "public and psychological wage" to its beneficiaries by virtue of authorizing the destruction of non-White people, their lands, and their relations as properly belonging to the "White world" (Du Bois [1940] 1983, 31, 45). Frantz Fanon (2007), in *The Wretched of the Earth*, better translated as "Damned of the Land,"

economic, psychic, social, and epistemic worlding structure, system, and program built out of installations of property, thingness, death, and the very idea of the human.

Generations of Indigenous scholars and activists have energized a field of critiques, debates, and concepts that help us maintain focus on the ways Whiteness is not just about bodies and persons but lays the groundwork for land theft and genocide, reactive projects to already existent Indigenous ontologies of being through and as land.[21] Postcolonial scholars have moved us to think more deeply about the secular liberal European entanglements with multicontinental colonialisms and the exclusions tied to One World imperial projects.[22] Contemporary abolition movements explicitly identify policing and incarceration as constitutive anti-Black geographies that require not only dismantling but generative new world building. Environmental justice, moreover, as a movement, a mode of resistance, and a way of knowing is grounded in the understanding that environmental violence is a manifestation of racism and colonialism.[23]

These understandings and the movements and scholars that made them are foundational. Thus, there is nothing novel in our saying that massive violence is underwritten by White Supremacy, even as the disciplines habitually deny this knowledge. Attention to FDWP and its world-destroying mechanisms is simply remembering what is already available to know.

---

describes White Supremacy as a racist colonial worlding built out of violence to land, psyche, and life that exists in hostile relation to other worlding projects. We learn from Baldwin (2021); Robinson (2020); Mills (1997); Williams ([1944] 2021); Harris (1993); and Wynter (2003), in their distinct elaborations, that White Supremacy is an economic, psychic, social, and epistemic worlding structure, system, and program built out of installations of property, thingness, death, and the very idea of the human.

21 For theorizations of racism and Indigenous erasure within the settler-colonial process, see Moreton-Robinson (2015); Coulthard (2014); and Nichols (2020).

22 For a few of the essential texts that have influenced our thinking on exclusions, see Lowe (2015); Mahmood (2011); Mehta (1999); and Asad (1993).

23 See Ruth Wilson Gilmore (2023) for a monumental assessment of these deadly dynamics as exposures to early mortality. On Indigenous understandings of environmental justice, we turn to the works of scholars such as McGregor (2009); White (2018); and Gilio-Whitaker (2020).

While foregrounding the influence of White Supremacy in planetary crisis projects of the North American university, we have found it crucial to notice the multiple forms of FDWP. Whiteness in its heterogeneity has many shifting manifestations and is not only North American, nor is it the only large violent formation at work on Earth. We still don't know our solar system, let alone adequately grasp these vast oppressive structures, even as we concretely wrestle with them minute by minute in predictable ways. We started describing FDWP in terms of the complicity of liberal environmental managerialism with White Supremacy at our universities, focusing on the second- and third-order grammars that don't speak about race explicitly but nonetheless enact the epistemologies of Whiteness. As we were writing this book, we were forced to grapple with rising fascisms and the mainstream celebration of explicitly racist White Supremacy projects surrounding the university, and increasingly inside them. This forced us to think harder about the ways Whiteness in North American universities has both illiberal and liberal forms (among others).

The illiberal form of FDWP invests in shoring up White patriarchal dominion over life and earth; it takes the formation of explicitly asserting a license to kill, rape, bomb, abandon, and take without consequence. This formation strives to be unregulated, to act with total dominion and without regard, to access anything with impunity as its entitlement. This formation wants to maintain a world in which it can destroy for its own desire without being called to account. Its very power is itself produced through its unapologetic enactment of violence and self-affirmative dismissal of anything but taxonomies of the White planet, a gratuitous disregard for any voice but its own. Its tell is its total disregard for everyone but itself. Its quotidian position is to hold everything hostage to its own engorgement—terra-breaking for profit and pleasure. This mode will not be called to order. It will not submit to regulation or law.[24] It would rather destroy than share a world.

---

24   See, for example, Anderson (2016) on how White rage continues to circumvent basic democratic norms and possibilities in the United States; and Main (2021) for a theorization of illiberalism.

Oil multinationals, which universities continue to collaborate with, exemplify this illiberal formation of FDWP.[25] In the project of their own persistence, exuberant pollution must continue with disregard, structuring a hyperviolence so great that it threatens to shatter their own footing. Rather than submit to a single regulation, ExxonMobil offers to pay a self-calculated carbon tax, but only if it is declared to have no liability for its ongoing violent history (including being a top-ten world-historical polluter and funding climate change denial campaigns for generations).[26] The US state provides other examples. The National Highway Traffic Safety Administration predicts a 7 degree Fahrenheit rise in the Earth's temperature, but its forecast serves only as a right-wing tactic to release emissions from regulatory responsibility for establishing lower car efficiency standards.[27] They argue there is no reason a company should submit to curbing their airborne violence, once the long-term unrolling of the effects of past and present plumes is already installed as a future baseline, especially when so many other institutions and countries are opting out too. Somehow a global pandemic that differentially affects those chronically exposed to pollution becomes the perfect opportunity to suspend environmental monitoring and regulation and to attack the EPA. Or consider that the United States refuses to be subjected to international law, vetoing UN resolutions that would seek to control its flow of arms, limit its violence, or account for genocide in Palestine. Nuclear nationalism, it is important to remember, is founded on extermanism, on holding hostage global populations via the threat of nuclear attack and accepting collective world-ending as a normalized domain of democratic politics. The International Criminal Court may issue judgments

<hr />

25 On fossil fuel companies' contributions to universities, see Westervelt (2023). On the effects of big oil money on science within the university, see Almond et al. (2022). On the convergence of the Anthropocene debates with the illiberalism of the Trump era, see Connolly (2019).

26 For a list of the top global polluters, see Political Economy Research Institute (2022).

27 During the first Trump administration there were several official efforts to mobilize global warming predictions and the COVID emergency to increase support for fossil fuel companies. See Eilperin et al. (2018) on deregulation efforts due to predictions of a 7 degree rise in global temperatures; McIntosh (2020) on efforts to suspend environmental monitoring during COVID; and Holden (2020) on the diversion of $3 billion in COVID aid money to fossil fuel companies.

about illegal acts of war, but it can do nothing to address national desires for elimination, or new technologies of land clearing, or the orchestration of international order through nuclear threat. To be unregulated—emotionally, physically, cognitively—is the pinnacle of illiberal White colonial masculinity. In other words, within this illiberal mode, destruction is an endless means to an end, a form of self-fashioning through violence foundational to settler colonialism.[28] When the White Planet's architectures for profit and impunity are threatened, it clenches harder, digs deeper, and doubles down on fossil fuels. Then, when it can no longer cajole or coerce the world to sacrifice itself for patriarchal Whiteness, FDWP flips a suicide switch to burn it all down. *Just try to pry this dead White planet out of our cold, dead hands.*

The liberal form of FDWP is equally familiar and can be found in many kinds of environmental sciences, but also in humanities and social science research that unquestioningly takes up its terms, from earth system to the Anthropocene. This liberal mode recognizes and regrets the destructive effects of its own logics, then proceeds to project its recognition into a technical salvage project.[29] This mode is recognizable in the North American versions of the Anthropocene. At its best, the liberal mode of the American Anthropocene uses claims to a universal humanity and a single world to point out the concatenation of environmental harms, without noting how the harms are caused by some more than others. Chickens, insects, and microplastics all come to matter but only at planetary scale. But the contradictions of inequality and its acceptance of a postulate of a One World universalized problem amenable to engineering solutions dog the project at its deepest level. The limits of its politics come to the fore, as it must leave all the logics of property, migrant detention, individualism, supply chains, war, and heteropatriarchy intact.

The liberal mode of FDWP may not directly name itself as concerned with Whiteness and may even posture as involved in an antiracist rescue mission. This liberal mode can manifest more opaquely, in the normal-

---

28 See Slotkin (1998, 2000) for detailed assessments of American mythologies of frontier violence as essential to the creation of a White national subject. See also Drinnon (1997); and Cole and Shulman (2019).

29 See, for example, Neocleous (2008) for an assessment of security as a liberal, bourgeois concept; and Masco (2006) on the insecurities inherent in nuclear nationalism.

ized practices, infrastructures, words, and calculations that seek to manage and engineer the One World. The North American university, we have argued, is deeply structured through these second- and third-order manifestations of FDWP and its possessive Whiteness, found in concepts like planet, earth systems, and species. These include the insistence on a world of inert objects, the pretense of objectivity, the reliance on extraction and surveillance as modes of knowing, and a commitment to mastery and mapping. The liberal mode of FDWP is characterized by the heroic technical fix that not only leaves the illiberal form of White Supremacy intact but celebrates White possession as the cure for White possession. This mode is so ubiquitous that choosing one example feels woefully inadequate, even if offering more starts to sound and feel like a deranged litany. There is the liberal project of carbon taxation, which offers a capitalist cure and leaves the fossil fuel order of extraction intact by offering work-arounds to monetize carbon sinks. The plastics and products industry strategically promotes recycling as the win-win solution to address the 45 million metric tons of plastics currently used annually in the United States, while the actual percentage of collected plastics recycled continues to decline to an all-time low, below 5 percent. We are told and tell ourselves that our careful collecting, washing, and sorting will address the problem even while 95 percent of what is put in recycling bins ends in the landfill and that the global production of plastics is projected to increase 300 percent by 2060.[30] Or consider the Harvard colleague who explained that geoengineering was in fact ethical and necessary as it would save many people, in contrast to ending fossil fuel capitalism, which would lead to massive starvation. Reifying the expert savior, he was unable to imagine that most people live despite capitalism, not because of it. This is the dominant mode of so much research about massive environmental violence and the primary preoccupation of this book. We know we would seem more rational, less excessive, if we only gave one example, but there are four of us, and so many of them and more every day. Liberal FDWP projects, even if we do not accept their terms, are terraforming endeavors.

This is why we are striving (but also stumbling) toward more robust vocabulary and methods for how to commit to making less hostile

---

30 See Greenpeace (2022) for an assessment of plastic recycling rates; and Materic et al. (2022) on the saturation of the global biosphere with microplastics.

worlds. Engineering, design, prototyping, demoing all promise that new conditions are constantly possible; they are terraforming activities that seek particular futures.[31] The limit of the liberal Anthropocene is perhaps most explicit in its technocratic projects, in geoengineering fantasies that seek to include the entire planet in its vast machines. Liberalism in FDWP gives us the figure of the geoengineer, who gets to control and call it rescue, who gets to kill without calling it killing, who gets to break things and call it protection.[32] In the savior dreamscape of FDWP, everything of the dominant order is replaced back in its original position. The issue is not the viability or desirability of altering the course of climate change. The issue is the exclusions that are always already encoded into the concept, the understanding of who or what is of value, of the risks to be accepted or ignored, of the futurities that are coded into each world-making/world-breaking action. These techno-imaginaries are plugged into terraformations with lasting material effects that also demonstrate a world built for the few against the many. Worrying about environmental displacement, the liberal FDWP, which presents itself as the solution to itself, shores up the nation-state and criminalizes poverty as a means to steward the earth systems of the planet, and repurposes racist policing, prisons, and deadly force as a means to a hygienic end. Liberalism, in our view, is bound up in an inhuman property regime founded in exclusion. Another way to say this is that liberalism has always been a terraforming enterprise, transforming land and bodies through possession and dispossession. As university-based researchers riding the Intergalactic Bummer Train, we see our actions too as always already noninnocently terraforming, shifting relations and material conditions, even if at small scales.

---

31 On prototyping, see Halpern and Günel (2017); on the expansiveness of design concepts and practices in environmental thinking, see Günel (2019) and Busbea (2020).

32 Vice President Lyndon Johnson (1962) personified this view while announcing the launch of one of the first US weather satellites, stating that ultimately such technology would "permit man to determine the worlds' cloud layer and ultimately to control the weather," concluding that "he who controls the weather will control the world." For an iconic technicolor presentation of weather control imaginaries from this era, see Kimball (1959), a filmic Walt Disney coproduction with the US Department of Defense. It was part of the *Tomorrowland* television series, which offered television viewers "science factual" depictions of what life would be like in space, on the moon, and on Mars, broadcast just as NASA was being formed.

Though we are characterizing these two sides of Fear of a Dead White Planet as liberal and illiberal, this does not mean that they are always in opposition. Liberal rationalism and the claim to absolute dominion that is the toxic formation of patriarchal Whiteness have long worked in concert. Disaster capital, climate profiteering, counterterror, and venture capital embrace destruction as the way to securing vast fortunes, differentially expanding White possession and surplus value all the time. Liberal FDWP then proposes to save the day with a fix that does not have to call too much into question.

When Whiteness is dying, it wants to take the whole world with it—or at the very least destroy other worlds. It gets dangerous when Whiteness is scared. Whiteness doesn't want there to be a world without itself. It prefers a suicide pact. It chants "Guns, God, and Oil." FDWP is composed of an inability to imagine or allow a vision of a world without it—despite the many coexisting worlds that are already here, and more to be made. FDWP seeks a monopoly on life and death, ignoring the multiplicity of worlds or the universe of harms it enables. However, it is in fact surrounded by alternatives and sets the vastness of the alter-worlds against itself.

Letting things die

Even learning to die, learning to grieve, avow, and live gently in the losses of the environmental end-times is a Whiteness salvage project, which is also a project of enrollment.[33] "Let's all learn to die together in the wasteland wrought by human hubris—beyond all our differences, we are in it together on this beautiful blue marble." Don't make a fuss, everyone's doing it—including everyone that colonial capitalism and racist policing have been holding hostage and have already been killing. The *we* evoked here—in climate emergency, the pandemic, the financial meltdown, the war on terror, immigration panic, police violence, on and on—is the ruse of FDWP, refusing to see the vastly unequal distributions of violence.

No. Actually, hell no! We say let FDWP die. It was mostly interested in saving itself anyway. In fact, it has constantly drowned others, and it

---

33 See, for example, Scranton (2015); and for a critique of the Whiteness of the Anthropocene, see Yusoff (2018). On terminality and twenty-first-century technoscientific immortality projects, see Farman (2017, 2020). See also Evans and Reid (2014) for an important critique of the concept of resilience as a mode of abandonment.

is willing to bring it all down to save itself. We can hold the things, relations, people, and ways of being that we hold dear without working through these FDWP optics, with its tools, or at its scales.

But any rebirth is necessarily partial. It has inheritances. FDWP infuses material infrastructures and epistemological practices to such a degree that even as we try to let it die, in an attempt to reduce its ongoing harm, it nonetheless conditions worlds. Rebirth is not starting anew—that is not possible—but aftermaths do hold other sensibilities, other methods. Other worlds continue to persist. While some of the same tools will be used, they hold other potentials. These tools were already stolen and can be taken back and repurposed again.

If North American planetary conceptions of crisis are a mirror of White Supremacy and racial capitalism, then we can approach them as a marvelous foil for revealing the liberal and colonial installations formatting so much of the scholarly critical enterprise. We can study what FDWP desires—how it maneuvers to the universal or where it responsibilizes the individual, where it seeks conceptual veins to tap, what it crushes, when it wants to do a deep dive. Attend to the gaps in its attention and the silences in its discourse. Following the crisis concepts conveniently pulls one into the terrain of eclipsed worlds and gaslighting. It is possible to watch the new donor-driven disciplinary hierarchies being installed in the name of a universalized crisis: Cambridge's degree in Anthropocene studies, Stanford's new School of Earth, Energy, and Environmental Sciences,[34] Princeton's new neighborhood that conjoins environment and engineering,[35] Columbia's new Climate School, or Chicago's climate

---

34 Stanford established the Doerr School of Sustainability in 2022 to "power excellence across all areas of scholarship that, together, are crucial for advancing the long term prosperity of the planet"; see "Our Impact," Doer School of Sustainability, https://sustainability.stanford.edu/our-impact. This school merges earth and planetary sciences, civil and environmental engineering, and, as of this writing, has no faculty from the humanities or social sciences.

35 In 2022, Princeton University inaugurated the High Meadows Environmental Institute, devoted to "understanding the Earth as a complex system influenced by human activities, and inform[ing] solutions to local and global challenges by conducting groundbreaking research across disciplines and by preparing future leaders in diverse fields to impact a world increasingly shaped by climate change"; see High Meadows Environmental Institute (2024). The institute is part of the new Environmental Studies and School of Engineering and Applied Science campus under construction at Princeton. In 2023, the University of Chicago announced the creation of a new

system engineering initiative. Academic hunger games identify control of FDWP aims as the hot project, not harm reduction or the pursuit of alterlives. Humanities and social science scholars have charged to the gold rush of the Anthropocene, and to the COVID-19 pandemic, and before that to terrorism, and before that to the population bomb, creating a scramble for conceptual territory and legitimacy geared to the measure of a fully corrupted White planet formation. Environmental humanities remains a profoundly White space in most universities, as does the geosciences, both deploying a species-scale humanity as the only plausible organizing frame in these emergency conditions. Remember that concepts of the planetary are important precisely because they do this work—scripting the terms for habitability and responsibility in powerful ways that also erase their political implications.

The matrix of the FDWP project is so dense that it installs conceptual blockages. Here, we could pause to ask what forms of violence are unrecognizable in the scope of environmental research? Why is it that selling cigarettes on the street could lead to being killed by police but creating a pollution stream so vast that it generates asthma deaths around the globe does not merit any liability? How is it that we have regulatory attention to the consequences of a malfunctioning toaster but not for the malfunctioning national security state, which breaks international law, killing and displacing on the scale of the hundreds of thousands? Research that reproduces FDWP crucially does not enact accountability or seek to reduce violence; it simply describes a historical and material condition in minutia as a necessary outcome. This is why we start by describing the problem differently, seeking to maintain lines of accountability. What makes a world habitable and for whom?

FDWP is overwhelming, and yet there are so many of us here in and against this hostile and deadly place. While we need the critical diagnosis, we can't stop there and get stuck in its vortex. The whole point of naming FDWP is to do something else, to let it die, to dismantle it and look for other modes of engagement. Let's provincialize FDWP. Many of us know all these histories and patterns all too well. This is not at all the first time that it is suggested that the university is built out of hostile worlds, that in order to do research, to study, the only way forward is to

---

Climate and Sustainable Growth Institute and devoted its first act to the novel field of climate system engineering—creating a first-of-its-kind dedicated geoengineering department.

say no, to refuse the invitation to be disciplined, to find some friends, remember your responsibilities, and start activating another way, attuning to different potentials. Other worlds are not only possible, many of us are already from them; others are headed to them. We are making and maintaining other worlds all the time. We are always, already intergalactic.

Dismantling and abolishing some things also means cultivating and protecting the already otherwises that are still here and embracing the speculative project of making better worlds yet to come. What has to be undone, stopped, dismantled to make room for other ways of being and doing? FDWP may build itself out of death, but that does not mean that death itself is a violence. Stopping some things is not the end but a start. Stopping is an action; it is a doing. Stopping some things—racist terms, universalizing ranking schemes, nonconsensual data extraction—is also about opening up space for something else. Fanon called this "inserting invention into existence."[36] If the conception of the human is grounded in anti-Blackness and colonial violence, then it must be dismantled, and at the same time new modes of being fostered, perhaps a new humanity, perhaps something else.

There are many other ways of being and doing. If you are in North America, there are Indigenous orders at work that call you into responsibility. If you are in diaspora, you carry with you the land relations of other places into this opening. Other people are already making refusals for you to learn from and support. There is more to respond to than FDWP can ever imagine. To get to this generative work of participating in the making of less hostile worlds, we have found it crucial to begin with undoing violent concepts, practices, and objects, thereby helping to make room for a more habitable something else. Undoing can be a positive project.

Where do we go from here? What is the next stop on the Intergalactic Bummer Train? Nowhere and somewhere else. As we have wrestled with FDWP from our locations inside universities, we have come to some propositions that now guide how we have tried to begin doing research differently. Rather than entertaining the fantasy of escape, we recognize that one is in the middle, where one already is, while looking for portals into other histories, other projects, pasts, presents, and futures that have been here all along, rippling on the edge. Too often the disciplines posit

---

36   Fanon (2008, 204). For theorizations calling for a new definition of the human, see Wynter (2003) and McKittrick (2015).

alternatives only as a glimmer in the future, which is preposterous. We are already living in the radical future of FDWP and know how that ends. Let's refuse that demand and look for other invitations. The past is dense with other ways of being and doing. And so is the present. And, moreover, there are many pasts and nows. So let's refuse the one-planet story as best we can.

Here in and against the university, let's experiment with working the tensions, taking a step into something else, not waiting for the better moment to arrive or the end to come. Instead: What can you dismantle? What can you break? What can you lean on? What can you block?

Stumbles, incomplete gestures, partial dismantlements, noninnocent refusals, repeated unlearnings, stretching desires, neologisms that don't translate, equipment held together with duct tape and rubber bands—failure here is part of how to proceed. You might have a feeling of vertigo—we often do—because you have detached from what promised to give you mastery. Let mastery crumble, stay unsettled, do not accept the order and calmness of rationality, the hubris of sure-footedness, for there is possibility and honesty in this vertiginous space. The rant, the slow thought, the paranoid insight, the love letter, the awkward gesture, the joke, the hustle, the burn—let's stumble.

A stumble is also a step.

Besides, when here, remember you are not alone. Working in joined-up experiments is where we started, not knowing where we were headed. That is a big part of why this book is odd. Who and what moves you? What compels, holds, and repels you? What are you grieving? What holds you up? What are you transformed into here? And now here? What sensibilities and potentials come from becoming with others rather than dominating them? What forms of reciprocity and recognition come from toppling over, staggering into, slipping through? These objects, forms, and forces that trip us up are possible allies but not innocent ones. They have the potential not just to disrupt the order and hubris of FDWP that constantly recruits all of us but instead might enable refusals and the creation of other sensibilities.

We assume you are here reading this because you already care about collective conditions, surviving hostile spaces, and the problems of living well with futurity—thank you! We assume you too are experimenting with study and joined-up tactics. So what does it mean to meet up on the Intergalactic Bummer Train now and refuse any assigned seats? Some may be tempted to rush the conductor car, but we don't want to reinstall a master operator. Some may be tempted to rush the engine room, but we don't want to reinstall a master engineer. Others may want to rebuild the train, but we don't want to reinstall the one-train world, with its linear tracks, let alone accelerate to high-speed bullet train status (like *Snowpiercer*). Meeting at this strange problem space, there is no singular *we* here, but the world of many worlds. So what can a train become for a multitude? How to not only disassemble FDWP but make something else? How can we not merely salvage but reimagine?

Let's start bringing our shovels, our sewing machines, our songs, our sense, our hearts, and the many tools we know you may have that we cannot even imagine. Let's start doing the joined-up work of figuring out how to terraform out of this bummer train for greater respect and habitability. By studying its copper wire and circuit boards and fabrics, its steam and spaces, let's speed toward multiworld repurposing, without prescribing for each other a master plan. There is hot soup and coffee waiting in what used to be the dining car. You might want to check out who else is there.

2.2 SOME PROPOSITIONS

All the work marshaling the evidence of a damage-filled planet has yet to mobilize a shift in how capitalism, colonialism, or racism work. Yet there persists a common academic assumption that if you can articulate a measurable condition, you will get the right politics and levers to respond to it. Counting something makes it real, gives one a quantity to manage, elements on a dashboard to adjust.

But the purpose of studying how violence becomes installed into surrounds and substrates—inadequately named by words such as *environment*—is not to make yet more measures. Our aim is to make less hostile conditions; to embrace the practice of attempting to make the worlds we are in more habitable and to dismantle the normalized necropolitics of FDWP. To participate in this kind of world-making—or what we call terraforming—means emphasizing the materiality and nonhuman dimensions and forces in such a practice at the modest level of study. The bummer, alas, is that good tools and methods for terraforming will not come from the planet-bound disciplines, and an individual academic alone cannot hope to manifest an equally scaled project in response. We need help, community, political movements, and permission to work differently. This is why we keep attempting to rearrange our modes of study and conviviality so that our small way of terraforming can grow. Practices come from other joining-ups, of thinking, working, and being together in old and new ways. There is incredible power in this realization, and the North American university abounds with spaces and opportunities to subtly terraform for better conditions.

FDWP is not the only terraforming force. We live in and participate in terraformations that can be otherwise. How can we attune to the abundant opportunities to alter conditions—to terraform for harm reduction and mutual support—in our own work?

Here is a set of propositions that helped us in gathering ourselves and our terraforming practices for this book during our collaboration. They are not a solution or a fix, but a stumble in a direction. Grappling with their contradictory entailments helps us to locate different starting points while attending to the histories and implications of our concepts and methods. These are only suggestions, and there could be many more such propositions—we hope to learn from your own list of propositions. These emerged from our efforts to unlearn and dismantle FDWP in our own habits, while improving habitability for teaching, thinking, and researching with others.

There is no starting point, but we need to start somewhere

Only the constitutively lucky feel the apocalypse as forthcoming, merely a future potential.[37] If you want to kill people, you will probably pick a starting point to justify your violence and erase everything else: October 7, January 6, September 11, July 4. The rest of us know that the violence is not forthcoming and is not out of nowhere: It has already been ongoing, and wherever we are, we are already somewhere in it, even if differently.

---

37 As Cedric Robinson (2020) argued, racial capitalism in the form of anti-Black slavery stole worlds away, turning people into cargo and building ongoing incarceration and plantation as material and spatial world orders. Colonialism ripped people from the land that constituted their kin, turning the living being of earth into property and genocide. For Indigenous people of the Americas, the apocalypse began in 1492. The world is fertile with the long trails of these hostile worlds and the survivances from them. The past is a constellation of starting points, repeated turns, ruptures, and possibilities.

Where we are, in the North American university, for example, is on particular Indigenous land. There are implications and responsibilities to being where you are that you may not know, but that are already here. These responsibilities are already calling you to an orientation that is before and beyond any laws of the settler state. Where you are situated—always somewhere—necessarily matters to your study. If all you do is everything in the most sanctioned ways, you are participating in keeping conditions as they are, and as a result, you are likely reproducing colonialism and Whiteness. At the same time, there is so much to do here, so much that needs working on; where you are leads to an open potential of starts. No one start is necessarily better than the other; step in any direction and there is work to do. Starting is simply about beginning not to be the conduit for the reproduction of FDWP. Starting is not about having the answer. Not about fixing something. It is about not waiting—not waiting for an innocent or perfect place to begin. You are already there.

The Charismatic Mega Concept (CMC) is a trap, but we still need concepts

What tools can unpack the One World and its powerful seductions? When the Anthropocene (or global health, or GDP, or counterterror) articulates a species-planet relation, its power derives from claiming ultimate scale in a way that eliminates the vast and different ways that people live, the multiple worlds that coexist and are in tension, and the wildly divergent impacts that communities have on collective conditions. The One World/one planet/one species move of planetary emergency thus stages the figurative ground for the heroic work of the geoengineer, who speaks and acts as a nonelected species representative, empowered by a salvation/savior sensibility. This is the danger we flagged with the Anthropocene Charismatic Mega Concept (CMC). But climate change is a CMC too. Researchers are surrounded by CMCs, large technical systems that animate only particular worlds into existence: globalization, macroeconomic productivity, world systems, national security.

We can now spin out analysis across scale from the nano to the orbital, and thick archives of facts can make clouds of correlation, a dense sphere of relations visualized through big data and high-throughput analysis. The CMC seems to tell you where you are, when you are, and how you are. But that is the trap. The clean periodizations, the gathering of all points into a single plane, the axis for comparison, the performance of precise location: These are all noninnocent. You are always already in the problem, already altered, already in the aftermath, trying to endure. The *when* of settler colonial violence, the *where* of anti-Black racism, the *who* of patriarchy, the *how* of contamination always exceeds, and precedes, the categorization.

The CMC allows you to maintain epistemic distance while feeling mastery. That feeling is a good gauge if you're wondering about whether your concept might be becoming or already is a CMC. Regardless, you are already in the problem, already in terraformed worlds. So what more processual, humble, situated concepts might invite you into the political methodological work of harm-reduced worlding?

Study will never be enough; orient to resolution

Study will never be enough. This sentence means many things to us. Let us untangle it. What is study in terraformations? We imagine that critical study in terraformations means not only a mapping but also a spacing—a making room for what infrastructures and systems thinking disavows. It means to search for modes and methods that disrupt and refuse the regularized installation of violence into our conditions even as they enable us to characterize regularities and patterned relations.[38] It means to join a project of long emancipation.[39] It means to study violent systems in ways that are unfaithful to their reproduction. Study for us has come to mean research that might generate alter-accountabilities to violence, that might contribute to less hostile conditions even if only scaled modestly—as well as to underscore the otherwise futures that emerge from such an alter-wise mode of working joined up across a multitude of challenges.[40]

The complicities within which we study are up close, personal, already in your head. For those of us that have been in universities for a long time, this may be especially true. The concepts are not just doing violence out there. They are injuring us too, perhaps more softly.[41] But

---

38 Our thinking on study is influenced by Harney and Moten (2013); Fanon (2008); Kelley (2003); la paperson (2017); Ahmed (2017); Dumit (2014); and hooks (1991).

39 For a key theorization of long emancipation, see Walcott (2021).

40 On joined-up thinking in environmental justice, see Di Chiro (2008); Agyeman et al. (2012); hooks (1986); and Reagon (1983).

41 For more on this formulation, see Harney and Moten (2013, 10).

maybe not only. In this cramped space of complexity and complicity, study means also inserting invention into existence, making concepts to take phenomena back, being responsible to animate pasts, keeping our eye on the long project of liberation, summoning together collaborations and coalitions that segregated worlds deny, including outside the university. Study can become self-changing and world-making, energized by a commitment to reducing the violence of existing conditions in solidarities. To put it directly: What forms of political resolve and connection can enable harm-reduced worldings?

Understanding complexity is a matter not only of optics and precision; it also reveals what you care about, what is messing with you, what binds you are caught in, what violences need calling out, and how pain matters to the constitution of your study. In this way, we offer resolution not as a term referring to a scale of granularity of data or view, but as an invitation to an ongoing processual method of ethical and political resolve. Resolution as resolve is about inhabiting and obligating to the legacies, politics, movements, communities, and ethics of somewhere. How are you obligated to the specific Indigenous legal order of the land you study on? How are you obligated to defend a terraform? How are you obligated to consent or refusals? How are you joined up with abolition? Such resolve may yield commitments in scale; that is, the scaling of your study or attention: Do you have responsibilities to this tree, this architecture, this taxonomy? How do you fulfill and renew such responsibilities? Resolve in this key makes the question of scales and optics a question of commitments and obligations to people, beings, and forces.

"Study will never be enough" is not sighed in resignation, nor is it thrown like a weapon. It marks the limits of the modes of study that call themselves research and hails others to come.

Don't trust your facts, but don't trust your objects either

Here is another parable of untrustworthy objects more intimate than a planet: Biologists announced in 2018 the discovery of the largest organ in the human body—the interstitium.[42] The interstitium is the fluid in a body's connective tissue, flowing within and between cells and tissues.

---

42   For the announcement of the interstitium, see Benias et al. (2018); for more detail on its biomedical implications, see Rettner (2018).

It is 60 percent water. Enabling immune response and linking organs, it has resisted identification because it is a domain of flow and transport, of connection and conductivity, rather than discrete objecthood.

The acknowledgment of the interstitium was no mere addition to the catalog of organs. It was an event that signaled the end of a particular regime of certainty in knowing what organs we have and what organs are in the human body. It raises questions about what an organ even is. How did the organ come to be the organizational cut anyway, outweighing attention to connection and flow? What should we make of an approach to the human body that misses so basic an aspect of a human being? When what was thought only as the background—that which is between organs—comes to be recognized belatedly as actually the largest organ of them all, it reveals the extensiveness of the problem of starting with agreed-upon objects. The moral of this story: Researchers do not already know the objects that make up the world. The given starting and ending places are not ensured. To wrestle with such foundational unknowing and uncertainty requires more than epistemological relativity; it asks of us an ongoing practice of unsettling.

One reason we cannot fully trust objects is that so many emerge out of the same systems of instrumentalist knowing that created planetary-scaled problems in the first place. That is, many of the core disciplinary objects of environmental concern—from chemicals to climate to technology—that are mobilized to confront or describe conditions of ambient violence are themselves concept materializations of those very systems. Moreover, many of them are presented as if they are contained entities, separated rather than emerging from relations. Objecthood as a status, we strive to remember, is an epistemic way of rendering a world filled with nonalive being; that is, a world of dead things. Crucially, most other worldings are not composed of the object/subject relation of the One World.

We need a healthy suspicion of quotidian ways of knowing and doing, such as assuming we live in a world of objects. This is not a paranoid or dismissive practice but rather a principled effort to think about how categories are world-making, about how our modes of engagement and inquiry function given the limited sensory abilities of experts. This mode might seem defeatist and cynical. But it's actually awesome and social. And moreover, there have been people in the university saying this all along. People in the cracks have been working to make space for other kinds of study, for study toward the project of living in and beyond violent worlds, in worlds that are not objectifying. If so many of the objects are wrong, then the space for study is enormous. It is abundant potential, in fact. There is so much yet to be said. So much to dream. Where you already are, in the intergalactic interstices, it is already there.

Yes, one can still choose the wrong object, grab the wrong brain for your monster, and need to start again. We have restarted this project so many times! So what does one trust, or better yet, what responsibilities might one resolve to trust as guides toward the generation of better ways of being and knowing? Distrust is never enough. There is the ethico-political-methodological stance of standing with beings and people against violence. The accomplices are not abstract.

Terraforming is not in the future; it is now

Perhaps one of the most explicit forms of FDWP terraforming today is geoengineering, a conscious effort to change atmospheric chemistry—to inject particles into the upper stratosphere to block sunlight, create a planetary shield in outer space, or build a mechanism to extract carbon from the atmosphere to put it back underground.[43] It is a terraforming techno-fix that appeals to those committed to petrochemical capitalism and the fantasy of endless profit. The cruel optimism of geoengineering appeals to those already embedded in a violent order that they refuse to change.[44] Geoengineering, as a planetary nonconsensual experiment in terraforming, promises to help those who need to keep putting roofs on their multiple homes (blown off by intensifying storms) or who are worried about how the next flood or ocean rise could take out their waterfront abodes. Geoengineering is presented as a futurological commitment, a way of handling a worst-case outcome—it has been rebranded by the CIA as "climate modification" to make it more palatable.[45] Some advocates of accelerated research on geoengineering talk about it as a "break glass in case

---

43 For an overview of geoengineering concepts, see Caldeira et al. (2013); for an appraisal of the problems attached to various schemes, see Royal Society (2009).

44 "Cruel optimism" is Lauren Berlant's (2011) term for psychological and affective investments in social promises that are already failing, such as unending social progress, class ascension, and democratic order in the United States.

45 For the original reports, see National Research Council (2015a, 2015b). For a discussion of CIA involvement in the two NRC studies and official worries about unregulated or privatized geoengineering experiments, see Morton (2016, 354).

of emergency" plan, but they also assume it will be necessary—projecting a permanent commitment to atmospheric intervention. What these schemes rarely mention is that geoengineering as currently imagined would likely redistribute both cooling and harms—helping some locals manage an overheating atmosphere while relocating the storms, the floods, the heat waves to other places on the planet. At its core is the cybernetic dream of a planetary environmental system that could finally be brought under managerial control, tuned for the safety and comfort of those multinationals and empires willing to experiment on a planetary scale.[46] This is the pure dreamspace of FDWP—a fantasy of perfect environmental control and extraction without financial loss.

But geoengineering is just one version of terraforming, and a particularly violent one.[47] Terraforming officially names the cultivation of more habitable, less hostile worlds.[48] Worlds and the planet are not the

---

46　On the evolution of weather control schemes to planetary geoengineering, see Von Neumann (1955); Fleming (2011); Hamilton (2013); and Morton (2016).

47　Terraforming as a concept emerges most directly from science fiction, from works like Kim Stanley Robinson's Mars trilogy (1993, 1994, 1996). Dreams of large-scale earthworks with planetary effects have a long history; see, for example, Fred Hoyle's (1957) novel *The Black Cloud*, which details how a sentient extraterrestrial cloud first superheats the earthly atmosphere and then blocks the sun, radically cooling the earth, with monumental effects. Or consider J. G. Ballard's (1962) novel *The Drowned World*, about a London submerged due to ocean rise and a population in retreat to the poles; or his (1965) novel *The Drought*, imagining a world where radioactive fallout has created a film across the surface of all oceans, preventing water circulation and eliminating global rainfall; or his (1981) novel *Hello America*, about a failed geoengineering effort that leaves North America an all but abandoned desert for more than a century. N. K. Jemison's *Broken Earth* trilogy (2015, 2016, 2017) imagines a world so damaged by technical systems that new forms of life evolve that can move through rock and control earthquakes. The US Southwest appears in a number of major works on environmental destruction and technological failure: see Bruce Sterling's (1994) novel *Heavy Weather*, about storm chasers in an overheated future panhandle; or Paolo Bacigalupi's (2015) novel *The Water Knife*, about water wars between Arizona, Nevada, and California after the collapse of the Colorado River water compact. Bong Joon Ho's 2013 film *Snowpiercer* is explicitly about the catastrophic consequences of a failed geoengineering project, one that produces a new planetary ice age and leaves a small surviving population trapped on a runaway train (still organized, railcar by railcar, as a class system).

48　For an early technical overview of terraforming concepts, see Oberg (1981); see also Beech (2009). For a broad review of NASA environmental research in extreme environments, see Olson (2018); for a history of how early space missions influenced

same thing. Worlds come at many scales, and are the conditions of our living. We are in a world of many worlds.[49] Moreover, we are terraforming in small ways in already existing terraformations, and thus are always already caught in the cultivated conditions. Terraforming names the ways that we are already and always in relation to land and within worlds, making and remaking them in the daily acts of living and interacting with others. And this land has been here a long time. Land is here in the city and the seminar room. It is not territory, and not property, and not a resource. Land is already in a noncolonial relation to specific Indigenous peoples. It is a fulsome relation that makes all our beings possible. Terraforming is a relation to conditions that are caught up with colonialism but precede and exceed colonialism at the same time. Terraforming is what you are already participating in, just by being.

Terraformations are not in the possible future; they are already happening all the time. We think it is a good time to take back the concept of terraforming from FDWP, to resist its monumentalism and reimagine the practices that make it up.

environmental thinking in North America to create a new vision of Earth as a spaceship, see Hohler (2015).
49 See de la Cadena and Blaser (2018), as well as Povinelli (2001); de la Cadena (2015a); and Omura et al. (2018).

PART 3
MIDDLES

3.1 WHAT IS A MIDDLE?

Let's turn to some middles. What is a middle? When we say that we start in the middle, we mean we start from where we are, inside the problem, and at the same time approach objects for study not as things in themselves, nor as containers, nor as signals to decode, nor trains to ride, but as messy moments of some duration in an ongoing ordering of relations within material conditions. *Starting in the middle means inserting a pause in ongoing terraforming so as to register what beings, doings, and relations are in play*. Of course, what happens and is generated in terraforming exceeds any arraying of objects and concepts. To start with a paused object in the middle of its doings and becomings is one way to identify possible zones and agencies for altering or supporting the conditions of which it is a part. It is a way of participating in terraforming and attuning to conditions.

Starting in the middle means beginning with something as it is—in the somewheres where we meet it—and hence the middle is where you are too. That something might be a technology, a place, an event, a tool, or an edifice. It may also be a concept, a garment, a body, a fragment, or a feeling. Starting in the middle, wherever we start, also means knowing that objects as we find them there, as they present themselves to us, will be flawed, as will the initial habits and tools through which they become present to us or through which we strive to know them. Starting with objects as middles is also a way of dissolving them—not just undoing their coherence as independent things but also tracing their many other embedded vitalities and relations. That means the project of undoing is also one of dissolving and connecting them from their place in the middle, attempting to undo the work of FDWP.

Undoing objects in their relations is a helpful, even if insufficient, first step. In doing so, we are trying to cultivate a mode of study that is a kind of terraforming, that participates in breaking FDWP and moving toward alter-terraformations. But we are also trying to make possible and foster a different growth medium from which other forms and relations can emerge, even if we don't fully arrive there.

In other words, in this section of the book we are trying to experiment with building a craft of study, which needs joined-up and situated thinking. Our initial steps are not random, even as they are not fully mappable. In writing these middles, we have attempted to practice and enact the propositions we have described above and the impossible methods that we list at the end of the book. Our middles start with a question— What is an X?—and then strive to give something back. There are so many places to start, so many middles, so much to do! But let's flag our commitment and start trying.

3.2 WHAT IS LAND?

Black ash trees reach upward from the swampy spring ground. Beechnut trees attract skies thick with pigeons. The trail passes by patches of tamarack and willow, and then over into the sugar bush with colorful maples. A thick black gum oozes to the surface. When the colonizers try their first survey in 1833, the deep wet of the swamp is impassable and repels their attempt to impose the line of their grid.[1] There are already people

---

1 For a discussion of colonial survey and mapping operations in Canada, see Burr (2006, 21).

here who know this *mashkiig* well, who tend this *iskigamizigan*, who use this black ooze, who understand law, *inaakonigewin*, that wells up from the land.[2]

In 1849, Canada's first geological survey starts the process of detecting and enumerating the economic qualities of this land. Canada is only a province of British North America, in turn a colony of the British Crown. The survey involves three colonial surveyors and a chemist who will test the material samples collected. The purpose is not a mere mapping of terrains for knowledge's sake. The purpose is to identify economic ventures in the land, to pinpoint the location of gold and iron, uranium and cobalt. In the black ash swamp, on lot 9 of what the colonists called Enniskillin, the surveyors find petroleum in the form of surface "asphalt." Land is turned into a resource, then property. Land is turned into settler substance.[3] The findings of the survey are published in *Scobie's Canadian Almanac*, containing "full and authentic commercial, statistical, astronomical, departmental, ecclesiastical, educational, financial, and general information" to aid local settlers. But news of the black ooze also goes global, showcased at the Great Exhibition of 1851 in London.[4]

Soon the land speculators arrived, their arrival enabled by a chain of practices for turning land into property for extraction and control. In 1827, Treaty Number 29 asked Anishinaabe leaders to concede over 2 million acres of land in exchange for some $10 per person each year for their lifetime, the terms of which were added after the signatures. Four reserves were established for four Anishinaabe communities, and the 2 million acres were given over to the Canada Company to sell.

The land's sellability was, of course, not given. It was an effortful colonial enterprise. The conversion required laying a grid, measuring with chains, writing documents, establishing a records office, pushing the technology of the signature as a way to consent to land theft, and turning pieces of paper called contracts into permanent displacements from

---

2   Anishinaabemowin words are used here with the help of the Ojibwe People's Dictionary (Livesay and Nichols 2021).

3   Recognizing Sajdeep Soomal for the framing of "settler substance."

4   The archival research about this survey and the treaties involved was done in relation to the work of the Indigenous Environmental Data Justice Lab, particularly collaborative research with Vanessa Gray, Beze Gray, and Kristen Bos, at the University of Toronto's Technoscience Unit as part of the Land and Refinery project (Technoscience Research Unit 2019b).

3.1 The Great Exhibition of 1851 in London, where oil was exhibited as a new wonder. Illustration by Joseph Nash.

land, as well as establishing "colonization roads" (that carry that name to this day) to help ease the passage of settlers to their parcels. It took tremendous force, in other words, to turn land into property—force that actuated across numerous proximal and distant sites and interlocking techniques: a terraformation manifested via distributed world-breaking practices. Indigenous people across the Americas came to understand that the arrival of the surveyor was an act of violence. At the Red River, in the prairies, at what is now called Winnipeg, Métis and Cree people took up arms upon the arrival of the surveyors, setting off the Red River Rebellion and an era of embattled resistance against a colonial Canadian state bent on Indigenous elimination and land theft. Through these technologies, land was subjected to the violence of the White possessive, beginning with the survey that sought to turn land into a common global measure—the acre, the chain, the mile—that could be made fungible, exchangeable, monetized, and then speculated.[5] For settlers, land changed

_____

5 See Morton-Robinson (2015) on the White possessive. For a theorization of the violence inherent in the concept of property, see Nichols (2020); on the racial

from dirt to soil, defined by fence lines and deed papers, coordinates that fix markers in universal space, sealed on a contract: the singular official map at the universal coordinates N 49° 48' 55.656", W 97° 6' 40.932".[6] The very definition and purpose of the colonial state was to make sure that this paper seal would stick, that this meaning and coordinates of land would endure and that to question it would be to question settler sovereignty itself.

This was and is a vast terraforming operation seeking to erase Indigenous life and the already inherent sovereignty of land. Lured by the black ooze, land speculation was particularly intensive at Enniskillen. This is the site of one of the Earth's first commercial oil fields. The black ooze, pulled from the land, would become a prized settler colonial commodity, feeding the settlement's economic prowess, creating the terms for vast urbanization as well as a global hunt for more black ooze. At first, the oil was primarily used to make kerosene for lighting, offering a commercial alternative to the global whale fat industry (which quickly pushed many species toward extinction).[7] But by the end of the nineteenth century, large oil multinationals began to form, buying up oil fields and slapdash refineries around the globe, adding ships and pipelines and trains to the mix. Standard Oil, the prototypical monopoly company, quickly took notice and bought up all the refineries. Later, Standard Oil would become ExxonMobil, the globe's biggest oil company, infamous climate change denier and the notorious fourth biggest contributor to climate change–inducing greenhouse gasses.[8] The region around Enniskillen, meanwhile, would become Canada's petrochemical refining corridor, energizing settler colonialism into industrial form. The

---

colonial relations of property as well as its Indigenous rejection, see Bhandar (2018) and Dorries (2022).

6 On the creation of a standardized time, see Ogle (2015); on creation of a standardized map, see Rankin (2018). See also Karuka (2019) on how universal time and space were tied to both the transcontinental railroad and stock market trading.

7 For a detailed multispecies account of the whale trade and its impact on the geopolitics in and surrounding the Arctic, see Demuth (2019).

8 See the Carbon Accountability Institute (2019) for a list of the top twenty corporate contributors to global warming, as well as Popovich and Plumber (2021). On the project of global warming denial organized by big oil, see Oreskes and Conway (2011). For an analysis of the intensifying global warming projections by ExxonMobil scientists over several decades that were accompanied by the company's climate denialism campaign, see Supran et al. (2023).

black ooze would be refined; when refined, it would power industrial machines and vehicles, turn into plastics and fertilizers, and feed giant multinational companies, eventually inhabiting and powering the entirety of racial capitalist production, such that it seemed impossible to do anything without it. Perturbations followed, imperceptibly at first, as barrels of black ooze cracked apart in the heat of refinement and combustion into benzene, naphthalene, and sulfur dioxide. Then it became pollution as it pushed out into the air in plumes, into the river in spills, and came to be breathed into bodies and absorbed into roots. Now the perturbations register in accumulating conditions of cancer, asthma, and miscarriage, and distributed death, even as more black gum is drawn out from the land, separated, and burned. The colonial state gives its permission to pollute, to kill in particular ways and durations.

A historical and ongoing amalgam of measurements, techniques, documents, and machineries characterizes the terraformational project of turning the black ooze into the material animating fossil fuel capitalism and the grist of settler colonialism. This making of black ooze and swamp into property is not land itself, but the violent reduction and extraction of land from its specificity, liveliness, and relation.[9] The land becomes molecularized in this terraformation, something pulled apart and made dangerous, carried across continents in winds and water currents, and concentrated locally in the blue river, the turtles, the fish, and the territory of Aamjiwnaang First Nation and other fence line communities.[10] Understanding property as a violent terraformation, demonstrating how the transformation of black ooze into oil is an FDWP achievement, is just one step in undertaking an anticolonial commitment to land.

So what is land? Land is where we are writing from. One is always on land. There is not one answer, but many place-based, or place-thought, relations, particular to particular people and worldings.[11] When we do study, we are in relation to the land we are on, and not just the geoloca-

---

9 For an Indigenous Marxist approach to land relations and an anticolonial land theory of value, see Coulthard (2014).

10 For a community-led history of Chemical Valley and the development of the petrochemical industry as a form of colonialism on Aamjiwnaang First Nation territory, see Technoscience Research Unit (2019a). See also Misrach and Orff (2014) for a multimodal assessment of petrochemical infrastructures in Louisiana.

11 On Indigenous place-thought, see Watts (2013) and McGregor (2009). On Indigenous place-based research, see Tuck and McKenzie (2015), as well as Pasternak (2017).

tion you might be writing about. This sense of land is more than soil, dirt, mud, and earth.[12] For some of us who are part of Métis and Anishinaabe worldings, it is the fulsome sense of capital-L Land in all its relations. It is animate Land. It is living Land. It includes waters, airs, and all beings, spirit and ancestors, language and law.[13] This Land is incommensurate with property and object thinking. But this Land is not for everyone. In this book, written among our different positionalities living in North America and working in the North American university, we have agreed to approach little-l land as place-based condition and relations. We have agreed that land in this sense grounds worlds and worldings in its fulsomeness, even if differently, that land is a form of active ongoing being, that land relations are a necessary middle to start with. We have also agreed that not all worldings, people, or politics, such as diasporic and displaced communities, are primarily place-based in their worldings in this way, nor should they be, and that these differences and movements also matter to our practices of study and politics of terraforming, even as the material itineraries of black ooze and the obligations to anticolonialism join us up discrepantly in shared commitment.

For here in North America, the place we write from and gather in, to be in relation to land is also to be in relation to Indigenous jurisdiction in some way. It is to be in varying relationship to Indigenous land defense and the project of taking Land Back.[14] Orienting toward this decolonial relation to land requires commitment to recognizing and supporting these anticolonial and Indigenous stands. This early commercial oil field is on Anishinaabe Land. Some of us write this book and study on Anishinaabe Land. Anishinaabe pedagogies understand Land to be the first teacher and the source of natural law.[15] Knowledge and language do not derive solely from humans; they come from Land. When approached with the correct respect and teachings, Land demonstrates how beings are in obligation to one another. Moreover, in Métis and Anishinaabe le-

---

12   See Mathews (2022) for a consideration of multigenerational landscape coproductions in Italy; and Hobart (2022) for an assessment of the transformational and intertwined effects of colonization and refrigeration in Hawaii.

13   On capital-L Land as demarcating a proper name in Indigenous worldings, see Liboiron (2021).

14   On Land Back, see Pasternak et al. (2019).

15   On land pedagogy, see Simpson (2014).

gal orders, Land is the source of law.[16] Law derives from Land for people to follow. More than this, people, both human and nonhuman, are embodiments of Land. Land is a living beingness on another dimension that one has responsibilities to. As beings of, and not just on, Land, temporalities of becoming and generations pass through Land in all its relations. Reproduction does not need to be understood as something that only passes from parent to child in a species or a body. As beings of the land, future generations exist as potential, as those yet to come, in the Land, while past generations also return there.[17] Responsibilities to Land are not particularly about territory, though land defense and sovereignty often have that dimension. Responsibilities to Land are about specific obligations and relations that stretch forward and backward in time.

This version of Land is not universal. Again, it is not for you to take. It is specific to the Great Lakes and prairies, and to particular sets of peoples, though other people have related land relations. What we offer you in this book instead is this sense of nonobjectified land processually composed in specific conditions and relations that you are a part of. Importantly, this pedagogy of land is not about being in the bush, finding some unterraformed pure location. In this way of study, the city is land too, and so is the seminar room. Land is relationally present not only in our bodies but in the table, in the building materials made of petrochemicals, in the rare metals in your laptop, and thus in this text. We meet the land where it is, just as we start where we are at.[18]

"What is land?" does not have a single answer even if we are together obliged by our commitment to say that land is not property while fighting FDWP; even as we are committed to say that toxic pollution created from oil refining is land coercively weaponized; even while affirming Indigenous jurisdiction.

Terraformation as an analytic invites a foregrounding of anticolonial relations to land that emerge from where you study, from where you

16 On Indigenous law, see Borrows (2019).

17 For discussions of Anishinaabe, Haudenosaunee, and Métis understandings of reproduction as moving through land, and not merely bodies or species, see Cook (1997); Women's Earth Alliance and Native Youth Sexual Health Network (2016); and Shadaan and Murphy (2020).

18 On relations and obligations to Land, see Konsmo and Recollet (2019) and LaDuke (2017).

are at. Land isn't property; property isn't land. This anticolonial project won't be the same for everyone, even if you are still obligated to it.

3.3   WHAT IS A LUNG?

It turns out that baby rat lungs are good for making models of simplified worlds of exposure, breath, and damage for biomedicine and engineering. The tiny lungs, or rather their proxies—silicon volumes cast from the interiors of now dead neonatal rats raised in particulate exposure chambers—register through their shape and size the accumulated bodily violence of altered and inescapable atmospheres. The violence could be marked otherwise: Airs could be monitored at fence lines or in smokestacks. But the epistemic infrastructures are built to deny airborne violences as off-the-record externalities. Nothing to see here, keep moving. So bodies bear the burden of the damage and the burden of proof; bodies and their proxies must then bear the burden of signifying that damage. To begin to refuse the damage, might it be possible to begin to refuse the methods of damage-based research?

What else could one learn from a lung? Where might it take us?

A lung is air sacs within air sacs, in a breathing body. A lung is an abstraction to contain a multitude—not two sacs, nor five lobes, but a manifold of bronchioles and alveoli, channels and sacs. Channels from a mouth fork before branching to countless cul-de-sacs. Respiration is the drawing of air into the cul-de-sacs for an exchange, and thus lungs are also the porousness of cellular membranes, the capacities of particularly shaped molecules to attach to or recompose with molecules residing in the cul-de-sacs. Held in the tiny bays, the air draws carbon dioxide from the blood flowing in nearby capillaries, and gives it oxygen in return, before it expires. This reciprocal exchange is then joined by other nonconsensual interruptions by sulfur dioxide, or ultrafine particulate matter, or airborne bisphenol A.

A lung is a medium for an ongoing gas exchange. Its membranes and shifting volumes locate and condition an unavoidable tide; here, a breather's atmospheric surround flows toward a lessening pressure, sucked by a broadening cavity, is then held to trade before trickling or spitting out to make room for another draft. Interior contours draw, channel, and hold the outside in, their skins serving as a surface of deposit and as a membrane for exchange, and the texture of these membranes, as well as the

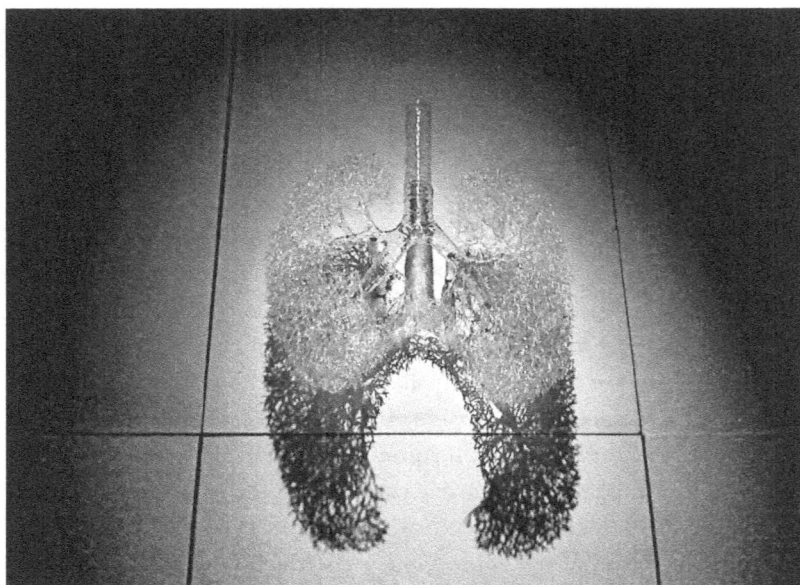

3.2 Annie Cattrell, *Capacity*, 2000–2001. Blown glass with lampwork. McManus Art Gallery and Museum, Dundee, Scotland.

shape and size of the volumes they form, all hinge on what comes in with the intimate expansive tide.

So breathing capacities are conditioned by the compositions and pressures of an atmospheric medium, where breathing bodies are contiguous harbors. Compositions and conditions shift as Mongolian north winds blow sands across vast regions and continents, smelters release fly ash into airstreams, forests in Canada ingest carbon forms and release oxygen, and oceans circulate the detritus of Fukushima's ghosts before they evaporate into new aerosols.[19] One could now fall prey to an atmospheric sublime or reaction formation thereof, endlessly exalting biogeochemical cycles, storm fronts, disrupted prevailing winds, or the driftiness of plumes. Or posit a singular humanity sharing a planetary predicament, a common future.

But we are with a lung. Not any lung—this particular lung. Before leaping to the expanse, ask where this lung lived and what this lung im-

---

19 On the politics of global atmospheric flows, see Nieder et al. (2018); Masco (2021); and Zee (2022). For a multispecies reconsideration of air, see Raffles (2010).

plies. Breathing is a necessary subjection; it pulls our attention back to a body, and from the inward into the palpable unknown of lungs encountering the molecular and particulate debris of violence. Breathing capacity is a confrontation between inside and outside, between this body here and the surround that keeps it alive, even as the same surround is killing. Lungs do not have a choice; we cannot separate them from conditions. We are caught breathing as the very condition for continuing to live.

Where does the lung live? What conditions become intimate in the lung? You have to breathe; or put another way, the choice not to breathe, to refuse breath, is of a particular disruptive form, bringing to an end the body-lung-condition of liveliness. Mostly, there is just being caught in breathing conditions not of our making, surrounds that blow hostilities, or scar membranes, or are set on strangulating. This being with conditions, and not just against them, is not a choice. Choke holds, choking haze, weaponized air—sometimes the strangulation is an act of physical crushing with hands; sometimes it is a dense concatenation of molecular harms. Breathing inhales nonnegotiable exposures to conditions beyond the subject.[20]

So we might say that lungs are responsive and absorptive bays in unevenly shared atmospheres. Lungs are the intimacy of life breathed, and at the same time expansive into the fulsomeness of atmospheres, surrounds, formed by actions both proximate and distant. Atmospheres hostile to breathers are not accidental side effects. These concentrated air conditions are the result of being the enemy of White Supremacy, of being criminalized for engaging in survival activity, of being a dumping ground for wastes.[21] The settler state weaponizes atmospheres as part of its land theft. It attempts to turn being into terra nullius open to the penetration of any harm without consequence. Whiteness distributes breath into zones where breath can be experienced as a smooth autonomic act not requiring attention and zones where breathing is an ongoing achievement, where it is a laboring act, where breath is made to continue against conditions. Fracking pads, oil refineries, and forest clearing are also about breath—they insist on participating in air, on the entitlement to disrupt the life-sustaining relations of atmosphere for the sake of piles

---

20 On the targeting of lungs and air, see Sharpe (2017); Kenner (2018); Sloterdijk (2009a, 2009b); Feigenbaum (2017); and Simmons (2017).
21 On survival practices in toxic environments, see Shapiro (2015); Agard-Jones (2012); and Murphy (2006, 2008).

of profit. These industrial projects participate in breath, in the lung, disrupting the possibilities of the lung, drawing those disruptions inside their machinery, converting life chances for the many into the accumulation of luxuriousness for the few. The expansiveness and enrollments of breath, which are also its intimacy, stretch back into the world-making, body-breaking histories of Whiteness, back into the carceral architecture of race, back into the clearing of the land and all its relations to make room for possession, back into the right to violate other beings without consequence, to choking without consequence. The now-dead baby rat is a companion in breathing. Atmospheres are not just flows of winds and clouds and chemicals; they are terraformed and terraforming distributions built out of the project of making the dead White planet. In breath, the dead are still present. Breath stretches backward, as we must breathe each other in, as we breathe the ancestors, as we breathe those that are yet to come. Breath requires cohabitation, corespiration, conspiring.[22]

It feels right to be suspicious of the conditions lungs are forced to reside with. The hostility of atmospheres made out of the material debris and force in the terraformation of FDWP are repeatedly disrupting inward and outward; the atmospheres are personal and at the same time sublime. Geoengineering promises to protect the lung by engaging a planetary-scale technosphere, not only terraforming but aero-forming, reiterating the industrial desires that created the problem to begin with. But we remember the lesson that the concepts and objects are not to be trusted. How to insert invention into breath? Beginning from a lung, or in a breath, the planetary dream becomes small. How to slow down, shrink, and stay with the lung when breathing is hard? Breathing is only ever done together. Who are you drawn close to by your breathing? What beings, what lungs, hold not only damage, but something else? The trees, the algae, the chloroplasts are here with breath. Lungs bring us to conviviality, to living with, and thus to problems of exchange, to reciprocal relations. Many lungs make the atmosphere. Study is impossible without lungs, and the relations they bring close into the where of living. Breathing is part of the labor of being together.

---

22 For a consideration of the conditions of possibility for solidarity for all breathers, human and otherwise, see Choy (2021).

A virus is a curious being—scientifically classified as neither alive nor dead. Viruses are also everywhere, both ancient and emerging, and some might even be extraterrestrial; they condition every living being and can be transferred via breath, touch, sex, or ingestion. A virus is a kind of parasite, living on the host but also capable of remaking that host in terms of both genomics and health. It is thus a powerful evolutionary force, one that has in its multispecies complexity organized life, from reproduction to intestinal processes to life span and pandemic. Viruses constitute a vast multitude—10^{33} in number by one calculation[23]—and can be found in every organism, in the oceans, in the upper atmosphere, and are able to jump species to create new forms, such as COVID-19, which is attributed to human contact with bats. Here again, the feebleness and limitation of separate interacting beings is undone. Zoonosis, the transmission of a disease from animal to human, assumes a clear distinction between humans and animals and so holds fast to the concept of species and all its troubles. How to understand a thing that can grow, evolve, colonize, respond to its different environments, and also wreak such havoc (COVID, anthrax, rabies, mad cow, West Nile, AIDS, Ebola)? For viruses are neither human nor animal, not bios or geos, but yet are consequential agents both of mass historical events and within us at the most intimate level, transforming bodily chemistries and biologies, confusing the very notion of *my* and *our* emotions, interests, desires. Viruses' qualities are already harnessed for recombinant vaccines, new formulations of pesticides, to insert genetic material into host cell's DNA for genetic engineer-

---

23   Felix Broecker and Karin Moelling (2019, 6) write, "The virosphere is the most successful reservoir of biological entities on our planet in terms of the numbers of particles, speed of replication growth rates, and sequence space. There are about 10^{33} viruses on our planet and they are present in every single existing species. . . . There is no living species without viruses! Viruses also occur freely in the oceans, in the soil, in clouds up to the stratosphere and higher, to at least 300 km in altitude. They populate the human intestine, birth canal, and the outside of the body as a protective layer against microbial populations. Microbes contain phages that are activated during stress conditions such as lack of nutrients, change in temperatures, lack of space and other changes of environmental conditions." See also Curtis A. Suttle (2005, 356), who writes that "viruses exist whenever life is found. They are a major cause of mortality, a driver of global geochemical cycles and a reservoir of the greatest genetic diversity on Earth."

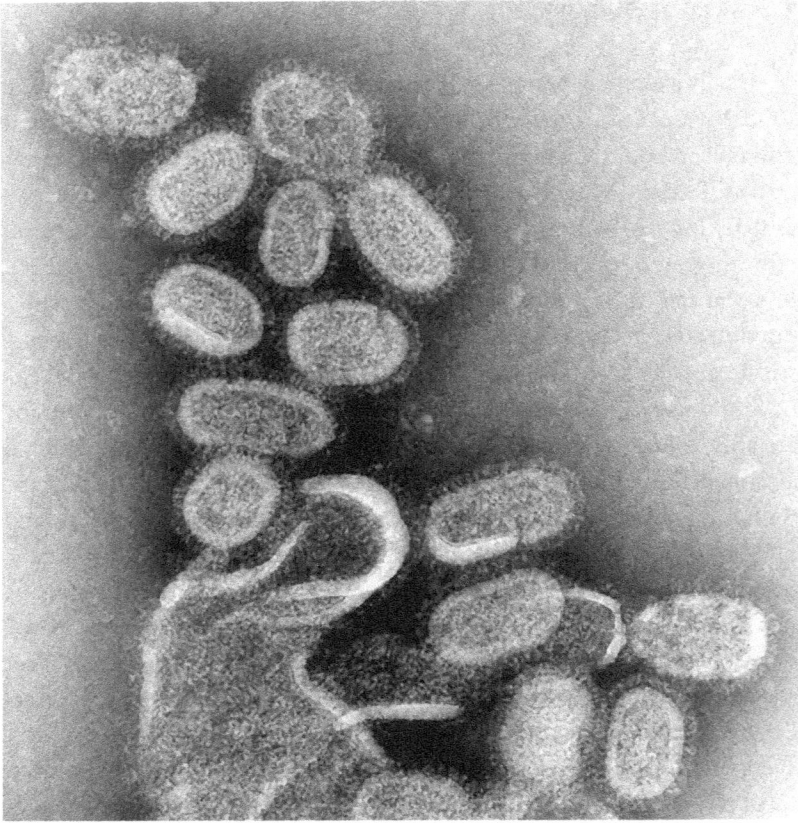

3.3 The 1918 "Spanish flu" influenza virus. Transmission electron microscopic image, 2005. CDC—US Centers for Disease Control and Prevention, Public Health Image Library, 8160.

ing, rendering them tools, into a managerial mode of relation. But viruses are terraformers far more than they are human manipulations, constantly remaking the terms of habitability—some are protective, others are activated by stress, some compel our behaviors, and some kill.[24] Because they cannot reproduce on their own, they are not considered living by biologists, and yet they can proliferate endlessly via the connectivity of beings.

24 For a discussion of how a virus is outside the living/dead dichotomy, see Povinelli (2016); and see Broecker and Moelling (2019, 57) on the world-making/world-breaking power of viruses, as well as Liang and Bushman (2021) and Zimmer (2015).

In the three short months from the official recognition in late December 2019 of a new SARS-CoV-2 in Wuhan, China, to the declaration of a global pandemic by the World Health Organization, a frightening, and yet familiar, array of connections and conditions was revealed by this novel terraforming virus, linking human-animal relations to global travel to broken health infrastructures to predictions of mass death on a planetary scale. Urban centers marginally aware of the new virus in January were shut down only a few weeks later—as stay-at-home protocols closed most businesses, confined people to their so-called households (assuming such a shelter existed), and reduced global airline travel by more than 90 percent. Nation-states conspired with the virus to reinstall the heteronormative household as a terraforming unit of necessity and shelter. You are in or you are out. What waits for you inside is your isolated problem. Racist and xenophobic reactions to a virus attributed to Wuhan, China, to Chinese nationals, immigrants, and citizens merged a contagious virus with contagious affect, revealing again the many ways people are connected across time and space, biology and psyche. If the interstitium revealed the largest organ in the human body as a zone of flow and transport, COVID-19 revealed the vulnerability of people to global contagions of disease, hostility, and misinformation.

In the United States, an economy suddenly frozen by fear of the virus, sets of interlocking relations were put on urgent display, showing how each person is made to embody a set of infrastructural relations and histories that predetermines vulnerability to death in ways that are also a map of race, class, gender, and age, of preexisting conditions (asthma and diabetes) and access to medical care or the absence of it. Death rates for African Americans in Washington, DC, by June 2020 were six times higher than for White residents, while the Navajo Nation was experiencing the highest per-capita infection rate in the United States. The virus exposed crowding, homelessness, lack of water, and employment status as indexes of foundational violences, the enduring achievements of dispossession and anti-Blackness. It sparked anti-Asian harassment, and ethno-national blaming. Racial capitalism enacted its deadly work, as the nation-state asserted itself as both force and failure. The COVID-19 virus produces a vicious viral pneumonia, degrading breathing and injuring lungs for those most affected, while leaving others as silent carriers without obvious symptoms, revealing how place, atmosphere, breathing, and community combine as destabilizing conditions. In this space of radical uncertainty about risk, people turned to masks (in short supply), to keeping far apart (when not pushed together by work or shelter), to staying at home (if such a place existed) to manage the amplifying pandemic. In doing so, the inequalities of shelter, labor, public health, and policing became ever more urgent life-and-death matters—forcing reappraisals of governance, crowds, family, hospitals, and emergency management. In short, COVID-19 as a material force revealed, in ways that could not be easily avoided, the terraforming discriminations within FDWP, within the nation-state as purveyor of death and life, and the specific embedded racisms and age-isms in a vast array of US institutions.

By spring 2020, the virus was an expanding planetary event even as it revealed each nation-state form as a specific configuration: COVID-19 restrictions on movement and industry led to an 8 percent reduction in global carbon emissions, even as the stay-at-home protocols kept safety workers away from petrochemical factories, leading to more spills from pipeline breakdowns, and the deforestation of the Amazon increased markedly due to a lack of state protection. The calls for more public health protections, always already embedded in practices of exclusion, were met by a formally illiberal Republican response, as the White president sought to eliminate environmental protections by arguing they were too expensive in a pandemic while offering emergency bailout funds immediately to big oil. Other Republican figures stated flatly that maintaining the economy was worth a vast human sacrifice of the aged, the poor, and the non-White.[25] The commitments of FDWP was on full display in public refusals to follow health protocols as a form of liberty and in calls for the sacrifice of unnamed others to support an economic status quo. Numbers took on new meaning, as metrics of infections, recoveries, and deaths became a primary way of understanding the pandemic, even as the constant sound of ambulance sirens revealed the amplifying emergency conditions as the weeks rolled by. The numbers, scary as they are, were also always false, a partial account collected only in hospitals, of a much larger illness and death rate. The numbers nonetheless informed biopolitical arguments about who is made to live versus who is let die.

If the illiberal response to COVID-19 was in full force with refusals of care and accusative storylines about the origins of the disease, the liberal

---

25　On how the COVID-19 pandemic effected carbon emissions, see LeQuere et al. (2020); for reviews of how the Trump administration cynically promoted big oil as a COVID provision, see Friedman (2020) and Volcovici (2020); and on Republican calls for mass death to protect the economy, see Beckett (2020).

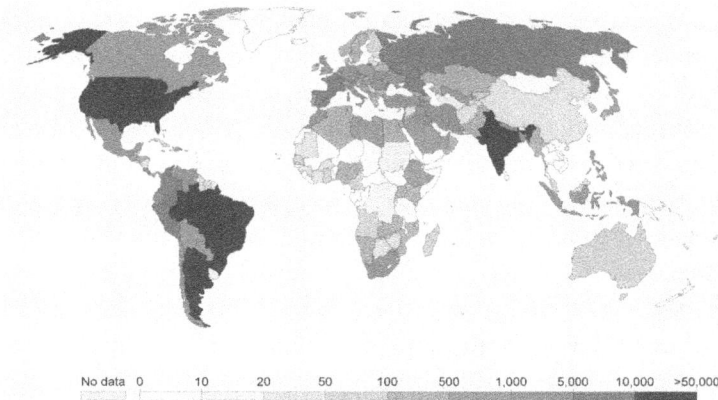

Daily new confirmed COVID-19 cases, Sep 16, 2020

Shown is the rolling 7-day average. The number of confirmed cases is lower than the number of actual cases; the main reason for that is limited testing.

No data 0   10   20   50   100   500   1,000   5,000   10,000   >50,000

3.4 Global COVID-19 cases, September 16, 2020. CDC—US Centers for Disease Control and Prevention.

limits of public health were also put on display as the US Centers for Disease Control and the World Health Organization revealed that decades of preparatory work on pandemic prevention and management were not nearly adequate to the disease. Shutting down public life seemed necessary because other options and infrastructures had not been built. Public health was both desperately needed and never enough in the pandemic, and still infused with the long social history of contagion and diseases as raced and classed domains, tied to eugenic campaigns and exterminations as well as hygiene, health, and care.[26] The twenty-first-century public health official in major cities—often a woman of color navigating the intense inequalities in places like Chicago, Toronto, and Oakland—emerged as the counterpoint to the FDWP politics of the White House, offering the primary expert guidance to scared and sick populations. In Chicago, the official "stay home, save lives" COVID-19 messaging rode on top of the profound health inequalities built up over decades—where be-

---

26   On US pandemic preparations before COVID, see Caduff (2015); on the emergence of a militarized logic of biosecurity after 2001, see Masco (2014); and for a historical appraisal of how infectious disease has been gendered, raced, and classed in the United States, see Wald (2008).

fore the pandemic life expectancy was thirty years shorter on the city's South Side (sixty) than on the wealthy North Lake Shore (ninety), the highest differential in the United States.[27]

The recursive and rebounding problem set of COVID-19, linking transmission vectors from animals to people to travel networks to cities and health care systems, to macro economies, to individual jobs, houses, family relations, care networks, and social media, put on display a global order connected both through its infrastructures and its abandonments, through its histories of care and gaps in basic services, which collectively determined which bodies receive emergency response and which were left on their own.[28] The post-2001 deployment of biosecurity to increase military budgets and to agitate citizens to fear their neighbors and enable more imperial war—with hundreds of billions of dollars in pentagon and contractor profits—has also been revealed as irrelevant to the terms of the pandemic except for the lost opportunities and resources. The related counterterror militarization of the police and border control linked the pressures of pandemic response to law-and-order policing, provoking an outraged public to erupt in every city over police killings—breaking pandemic safety rules to fight a more insidious and long-term form of anti-Blackness. The early 2000s US counterterror investment in biosecurity sought to militarize public health—choosing to fund a federal commitment to chemical air sensors that never worked over supporting emergency hospital care—just as the 2010 Affordable Care Act would ultimately be deemed a success for only leaving 27.5 million highly vulnerable poor people out of health coverage. As states closed borders, banned travel, and shut down economic activity, the effects of these choices were amplified, exaggerating already existing death worlds and suffering, leaving those most exposed to watch the stock market reach an all-time high while over 40 million applied for unemployment.[29] COVID-19 presented

---

27 See "Chicago: Public Health Statistics—Life Expectancy by Community Area—Historical," https://data.cityofchicago.org/widgets/qjr3-bm53. See also Lartey (2019).

28 Alexis Shotwell (2020) writes, "COVID-19 is a virus; it is also a relationship. Whether people live or die when they get sick depends on webs of social relations, the history of oppression carried in their bodies, what care is made available for them to receive and so much that we don't yet understand." See also Roberts (2017).

29 In March 2020 the US stock market dropped 30 percent as the lockdowns began, then started a wild climb, ultimately reaching a new all-time high in December 2021—double the COVID low.

an unavoidable lesson, not just on biological connectivity and shared respiration, but in the ongoing terraforming force of foundational violences.

What is COVID connectivity, and what are its perturbations? Thought through a virus and transmission vectors, connectivity figures as contagion, with the edges of individuals blurred through focus on shared airs and surfaces, making every person a potential threat to another, always potentially subject to being affected by another—with every body potentially (though not equally) vulnerable.

Yet the pandemic's connectivity should be thought both with and against this contagion. For while contagion offers a route around the liberal subject—a potent and ready-made critique of the illusion of a liberal subject living in a contained, only-human, body who can self-control and make rational choices for its own good—it achieves this via refurbished affective circuits of threat via contamination, where individuals and populations can only be experienced as dangers to one another. Health, then, by segregation. Threats to well-being experienced and managed as individuals, households, and populations rather than infrastructures of vulnerability to exposure and differential morbidity. At the same time, other connectivities have bloomed. Pods of assistance and redistributed resources offer alternative life support, following leadership and replicating models made through disability justice and queer mutual aid. Agitated states under lockdown commingle with anger at normalized racist and class-based exposures to death, amplifying energies for activating abolitionist futures. These pockets of aid, agitation, and activation are COVID's other connectivities, its alter-contagions, pointing to the limits of the present as well as the possibilities for terraforming worlds otherwise.

Within these structures, the university responds with corporate crisis modes, seeking to restore its baseline rather than rethink the ways that contagion operates, what it has revealed, and what it demands. Some researchers look for a vaccine while the rest move online—attempting to keep existing pedagogical logics firmly in place even as every single person is dislocated from the campus and in their lives in unprecedented ways. Have a problem, build an app—can I offer you a dashboard? The violence of the current global economic order is lost in the desperate efforts to restart it—to get the undergrads back in the dorms to take their virtual classes, to protect the endowment rather than use it to help people, to reassert the logics of disciplinary knowledge rather than rethinking the problems revealed by the pandemic. A lesser but important revelation of the moment is the structure of the university as a corporation rather than a commons, an entity that deals with emergency via disaster capitalism (restructuring, contract breaking, layoffs) rather than operating as an intentional terraforming force for harm-reducing worlds. Take note: Two years into the pandemic, austerity measures and a soaring stock market grew university endowments by more than 30 percent. Thus COVID = death/profits.[30] But, as the pandemic also inescapably demonstrated, we are all breathing together, all the time, whether one acknowledges it or not.

3.5   WHAT IS THINKING?

Academics are taught that we live in a political universe that privileges rationality and transparency; in such a universe, the ability to see things as they are enables collective judgments to be made based on the certainty of those appearances. This promissory note, however, rides on the distortions not only of mass mediation but of explicit, willful, expert logics of misdirection.

Deception has been fully weaponized in the petrochemical age and is now supercharged by digital media, competing global industries, and planetary-scale military visions. How was it that at one moment (coincidentally the year of a US invasion of Iraq), 70 percent of Americans understood that Saddam Hussein was involved in the 9/11 attacks, and sometime later understood this not to be true at all? When 99 percent of environmental scientists understand a warming climate to be an effect of greenhouse gas emissions—why is it that television news tells us that there is still a debate about the industrial contribution to a warming planet?[31] When the news rationalizes Israel's bombing of a Palestinian hospital because of a suspected military command center tunneled beneath it, breaking war norms—why is the news silent about the war crime bombing of some twenty more hospitals afterward? When a policeman kills an unarmed non-White citizen, why are there no readily available federal statistics on how often such violence has occurred,

---

30   For an assessment of university endowments during the pandemic, see Whitford (2022).
31   On the shifting public perception of the 9/11 attacks, see Milbank and Deane (2003); for an account of climate science denialism and media misdirection, see Oreskes (2004) and Lynas et al. (2021).

which police departments are repeat offenders, which bodies have been targeted, and so on?[32]

Psychological operations (psyops) is the military term for managing perceptions, or, as the Pentagon would put it, "controlling the information environment."[33] This term is about leverage, about controlling people for a specific advantage, manipulating feeling, attention, and thought all at once. From persuasive techniques to propaganda to covert action and strategic deception, the White Supremacist, settler colonial political universe of North America is structured by systemic acts of misdirection, willful silences, and enforced fabrications. In fact, it relies on them to maintain power and endure over time. It is common to hear, for example, that the United States is a democracy, founded on the idea that people are born equal; yet, from a practical point of view the US democratic experiment only started with the 1965 Voting Rights Act, which sought to enable one adult person–one vote for the first time. This fundamental democracy goal has been under assault ever since (denying most incarcerated people the franchise as well as those in US territories). It has also been the subject of advanced democracy reduction techniques organized under the psyops banner of defending against voter fraud.[34] In psyops, attentional capacities are resources that are extracted with increasing precision, targeted by people, corporations, state agencies, and emerging technologies whose continued ability to extract depends on public distraction, disorientation, or misdirection.

Psyops involve programs run by big oil, pharmaceutical companies, social media firms, political parties, the military—all attempting to naturalize the forms of violence that enable their endeavors, to craft a singular world that cannot be held to account, no matter the scale of injury. That scale can now involve the entire planet, rendered profitable to some via advanced forms of manipulation and perception control. People are sometimes invited, sometimes shoved, sometimes woken up to face understandings of collective life that are crafted, tactical, and purposeful:

---

32   In 2015 the *Washington Post* started a running database of people killed in the United States by police shootings: "Fatal Force," https://www.washingtonpost.com /graphics/investigations/police-shootings-database/.

33   For a formal psyops field manual, see, for example, US Army (2005).

34   For a detailed history of voter suppression in the United States, see Anderson (2018); see Browne (2015) on the origins of policing and surveillance in anti-Blackness.

designer perceptions and affective recruitments to render raced, gendered, classed, and ecological violences not as violence at all.

Psyops might be the most ubiquitous form of public knowledge today. When one asks within the space of the university how we come to know what we know, however, psyops are not usually factored into the assessment.[35]

Under such conditions, what is thinking? Under such conditions, this question animates not just a will to know but a tactic of learning (and at times, vital unlearning). Why do we think we know what anything is? How do we know where it starts and stops, if the starting places and received wisdom are actually part of the problem? Surrounded by information warfare and propaganda in a screaming 24/7 news cycle that runs on repetition and agitation, it can be challenging to trust our own thoughts, senses, and feelings.[36] I can see the flames, taste the smoke, sense the hostility, but the headlines seem to tell a different story. I walk past growing tent cities in my neighborhood, but I am told the numbers of people who are homeless are going down. Everyone can see the judicial nominee lying to get his job, but the old guard will only attest to his high moral quality, even as they plan the demise of civil rights, environmental protections, and corporate accountability with their future votes. The White President advises us not to trust our eyes or ears, changing narratives from minute to minute while taking children from parents at the border. Surrounded by such spectacular catastrophes in the space of FDWP, we are asked to embrace a performance of life as if nothing in particular is going on. It can be dizzying to parse the appropriate tenor of thought.

Are the thought-scapes that condition environmental violence best characterized as rational? What kind of transparency would render our academic disciplines nonviolent? The Intergovernmental Panel on Climate Change (IPCC) announces that even if all fossil fuel extraction stops today, climate change would still amplify to a catastrophic level, and in the very same moment the White presidential administration announces that car emissions no longer need to be regulated, yelling, "Drill, baby, drill!"[37] The university extols the virtues of diversity—it is a form of hu-

---

35 On colonial unknowing, see Vimalassery, Pegues, and Goldstein (2016); and on misinformation and conspiratorial reason, see Masco and Wedeen (2024).
36 See Orr (2006) for an assessment of panic as a mode of collective feeling; see also Berlant (2011) and Masco (2014).
37 See Intergovernmental Panel on Climate Change (2018) and Davenport (2018).

man capital after all—but the administrative meeting is a circle of White faces, surrounded by walls with portraits of genteel White men. The comportment is polite, but you know you are alone in a hostile place. The refinery catches fire; the pipeline does too; so does the forest; but the corporate press release reassures us—nothing going on here, keep moving. I saw the police kill her, but the report says she attacked first.

This form of thought is familiar. Psychologists and survivors of intimate violence have given this kind of psychic abuse a name: gaslighting. Gaslighting names a kind of mental manipulation, where conditions one can feel and sense are denied by the abuser. I am not hitting you—you asked for it. I am not doing anything wrong—you are the problem. The techniques of gaslighting have the effect of compelling one to distrust one's own sanity. They undermine the very force of thought. Gaslighting is a key tactic of FDWP, inventing the modes of racism it can deny.[38] We know this already, but it is unthinkable at the same time. Gaslighting and psyops often operate together, and it can be hard to know which way is up. War is presented as defensive, pollution protected as a clear skies project, voter suppression pursued as election integrity, and on it goes.

Gaslighting is nothing new, even as novel techniques proliferate through big data, social media, and corporate and government policy. Gaslighting is a core practice of the long arc of patriarchal violence. And it is a constituent tactic of FDWP: Whiteness erases itself even as it intensifies exposures to death materialized by the pervasive structuring of race into totalizing conditions. Settler colonialism fundamentally operates through gaslighting in its use of the legal conceit of terra nullius, empty land. Indigenous erasure ensures everything is open to stealing and destruction because no one is there, and yet we know the land is densely inhabited because we are part of the presence who have always been there. Gaslighting penetrates the ability to understand environmental violence by foreclosing on one's experience and perception. If one is measuring the pollutants, and the measurements detect nothing, then how can anyone else claim that something is actually happening? It is not just that the oil company denies and deflects—we have come to expect that. The permission-to-pollute systems of the settler colonial states of the United States and Canada are designed to gaslight, to deny the thinking about

---

38   For theorizations of gas lighting across race, culture, and environmental science, see Davis and Ernst (2019); Ruíz (2020); and Murphy (2021).

the violence that forms the conditions of one's existence and exposure to death, displaying a quintessential long-term psyops strategy.[39] What kind of thinking could render our academic disciplines less violent, or could steer research toward a political pragmatics of how we know what we know?

There are guides. People have worked hard to triangulate each other's senses of reality in order to diagnose the gaslighting maneuver and reveal the psyops programs. It is not so impossible to see through the sales pitch, the press release, the Twitter (or X) rant, the prime minister's tearful apology. Learning how to triangulate with one another about how to think about shared conditions within multiple worlds is a crucial thinking survival tactic. It requires places, friends, listening, conviviality, elders, mentors, music, poetry, grounding in your worldings.

Yet gaslighting runs deep in the university. It is built into the disciplines and their constitutive erasures, essential to their founding fictions and everyday tactics. It often colonizes affects and imaginaries, defining the parameters of what can be thought and the terms of professional discourse. Academic thinking can be a muddle of abusive gaslighting thought, slicing worlds into objects and domains, enabling exclusions and silences, and normalizing violence in the name of constituting expertise. Academic gaslighting can take the form of objects without relations, concepts without contestations, and datasets that deny their extractive conditions. Being in the university seems all too often to require as much or more unlearning as learning. Unlearning in order to keep thinking, to build trust in thinking together, to triangulate conditions.

Chemicals are a good example of a gaslighting object. Chemistry is a discipline primarily built by the chemical industry, skilled experts in misdirection. Why would its forms of thought work to confront environmental violence? Chemistry teaches that chemicals are small; they are molecular relations too tiny to sense with the body. The molecules must

---

39   For example, a recent study found that over 90 percent of Chevon's carbon offsets—a central part of its pledge to reduce to zero emissions by 2050—are worthless, constituting an elaborate fiction of corporate responsibility by the second-largest fossil fuel company in North America; see Corporate Accountability (2023). For a similar account of the vast sums that Shell Oil put into carbon offset schemes that also had no measurable effect on climate or the environment, constituting an elaborate greenwashing of oil profits as the planet continues to heat up, see Lawson and Greenfield (2023).

be studied one by one and are primarily bundles of physical properties amenable to engineering: atomic bonds, molecular weights, half-lives. What opens up if we triangulate together: What is a chemical? Living next to a refinery, caught in the plantation, downriver from the mine, we know that synthetic chemicals are massive. They nonconsensually penetrate and disrupt land and water, and change the expression of being. Synthetic chemicals have kin that need to be called to responsibility too: PCBs were made by Monsanto, as were Agent Orange and glyphosate. Seventy-one percent of the greenhouse gasses causing climate change is created by just one hundred corporations. We all have Monsanto and ExxonMobil materially constituting our bodies and exposures to death, yet in profoundly uneven densities.[40] Here is our invitation to drop the abusive lover, abandon the abusive concept, deny oxygen to the violent structure of thought. It takes life support in the forms of friends, community, and relations to make something else.

What is thinking, in the long haul of making other worlds? What is thinking in the intimacy of our joy and misery with one another here in these worlds we are making together with the help of our ancestors and the ones yet to come? Come, let's unlearn and make alter-thought in the corner, down in the basement, on the run. This is not terra nullius. Thinking moves across us, not out of us; we need each other. It is not contained in the neuro-connection, the computer algorithm, or the DNA test. Thinking is not a form of reason but a fabric, born beyond us but sutured into us, for making worlds, of which there are many, and we need many more. It can be composed in relations of resolve and responsibility, as a mode of survival and persistence, with obligations that stretch forward and backward in time. It can also be a refusal to allow violent conditions to continue without attention and pushback.

It is not easy. Exit is often not possible. Full exits never so. Escape from abuse can be a privilege. What kind of thinking can lend support or draw a new line when one is caught in, and must simultaneously be against, the violent condition; when one is in the problem and against it all at once. In the disciplines, our despair and our entire attentional space

---

40   On late industrialism as an era of exposures, see Fortun (2012); on glyphosate, Adams (2023); for a listing of the one hundred top greenhouse emitters, see CDP (2017); and for a consideration of what chemical exposures do to concepts of life and futurity, see Murphy (2021).

are mined for their capacities to keep the world as it is, not to remake it under less violent conditions.

This is when we need each other the most. If epistemic norms are a concatenation of infrastructures redirecting imagination, desire, dreaming, and emotions, then joined-up desiring and dreaming are essential to thinking toward something else. Why is it easier today to imagine the entire planet on fire than the end of racism? Why is it easier to commit to permanent war than imagine collective peace? Why is it easier to imagine that billionaires will save us rather than building an equitable society?

Or is it? Just try. Ask: What can be dismantled, burned, broken, or hacked in the terrain of thought with and against thought, of dream with and against dream, of desire with and against desire?

On the Intergalactic Bummer Train, we have been careening along on the tracks (the infrastructures, concepts, imaginaries, disciplines) made by empire as a terra-breaking counterformation to the world of many worlds. The imbricated material force of FDWP can seem like a runaway system. It keeps trying to lay our tracks, forcing us back to the same tired station of One Worldism end-times. With a little rocking or some quick switching, you might see more than you did. But it's hard to get somewhere else.

As we write these paragraphs, the bad times intensify. The planet is already heated to the benchmark 1.5 degrees Celsius, genocidal bombing is protected in the name of civilization, the global meeting to negotiate fossil fuel emissions reductions is being chaired by the president of one of the world's biggest oil companies, Trump is making a comeback, transpeople are innovatively criminalized, the news is a misinformation machine, and for many elites, geoengineering has become the only conceivable option for a superheating planet. In other words, FDWP's commitment to unregulated killing has only grown. Even with all the good company, speeding along on the Intergalactic Bummer Train generates a ferocious kind of trapped vertigo, a sense of the impossibility of change. This is a psyops effect of FDWP, one we have been attempting to disrupt through study.

If study is terraforming, what happens to the Intergalactic Bummer Train? Well, maybe it is time to take it apart too, to dismantle, burn, break, hack, and see what else can be made for better living conditions.

Why break the train? Well, for one, there are a lot of people we have met along the way, people trying to find a way through or a way toward something else, a sense that something of consequence might actually be possible. There are so many ways of joining up, doing solidarity, cobuilding. On the train, it is too easy to be rocked to sleep, to accept only the scheduled stops. It's too hard to connect between cars, too easy to miss what is beyond the tracks or on the other train speeding in the opposite direction. Most of the windows do not even open anyway—and right, the dining car food was never really that good. There are many better efforts creating harm-reducing and even desire-based collective conditions in other places, less freighted and stuffy than our train made with so many parts and pieces from the university. Empire's tracks will only take us down imperial routes. While the Intergalactic Bummer Train got us laughing to the world of many worlds, we don't need its prelaid tracks, devotion to universal speed, and singular direction. What is this train anyway? That problem space has now shifted: We are no longer just worried about overwhelming violent conditions reproduced in the university's environmental research; we are trying to know and be something else. The bummer train has already come off the tracks. We've been taking it apart along the way. But it is time to commit. When the wheels fell off in part 2, we never replaced them. A few cars are now on disappearing tracks, heading in different directions, some starting to look like space canoes.

Time to switch metaphors. Time to build something of our own to go somewhere else.

As terraformers, we orient to how conditions emerge and attend to the problem of always already being embedded within them, and thus shaped by them.

Our key concern in writing this book has been to think about collective conditions without reproducing the violence that informs them. Rather than starting with study and hoping to change the world, we have come to the ongoing commitment to worlding in a world of many worlds, working to make more habitable conditions and relations with and against the university and beyond. It is out of terraforming that other modes of study become possible.

Study alone is not enough. This is another way of saying that modes of study that imagine themselves isolated from their terraforming relations will not just be inadequate to environmental problems but also dangerous. To commit to terraforming study is to commit to the multiscalar and tangled responsibilities for altering and being altered by our relations and conditions. We emphasize the intergalactic nature of this prismatic project of terraforming, as the perturbations can come from far away, and you are always already in the mix as agent and recipient—and other worlds swirl around. Starting in the middle is also starting with terraforming.

If you are feeling blue or simply exhausted by our itinerary to date, we get it. The next section tries to articulate some methods and ways of being that acknowledge the desires we have for a form of study influencing conditions in relation to each other and our responsibilities. Kidding/ not kidding, we will call this *terraformatics*.

PART 4
TERRAFORMATICS

4.1 THE RESOLVE

In her 2015 book *The White Possessive*, Quandamooka scholar Aileen Moreton-Robinson quotes her uncle Denis Morton that "the problem with White people is that they think and behave like they own everything." The totalizing logics of White possession are a vast effort, couched in languages of quantification, emancipation, progress, reason, improvement, and vanguardism, while fraught with fear and motivated by greed. These logics are not simply ideas. As we tried to show, they are installed through myriad world-building practices of engineering, collecting, surveying, demarcating, containing, measuring, communicating, wiring, naming. These are the second- and third-order manifestations of FDWP that we have been wrestling with through this book as part of our effort to refuse planetary crisis and environment as frames for our politics and research.

For us, orienting toward terra has been a commitment to place-based, obligated nonuniversal study in a world of many worlds. It is not coincidental that we say *terra*, because anticolonial struggle in North America begins with defending a preexisting and continuing Indigenous relation to Land incommensurate with property and fundamental to specific Indigenous peoples. In discussing what it means to start with terra and Land in a settler state founded on dispossession, anti-Blackness, and immigrant exclusion, we have wrestled with our different positionalities. This book's attempt to think with and against academic work to describe place-based conditions should not be read as commensurate with the relations to Land that specific Indigenous peoples have across the Americas. Instead, it is a prompt to resolve one's political and obligatory relations to place and land as a crucial starting point of anti-FDWP research. We also carefully use *place-based* with an awareness that rootedness is not a norm and that places are constituted in fluids, transits, migrations, diasporas, and displacements, forced and otherwise, across these various positionalities.

What is terraformatics? This is what we jokingly call our approach, our resolve. It sounds grand and absurd and makes us laugh. Though it might be more accurately and ridiculously called *post-planetary intergalactic terracoformatics*. The method is how we keep aspiring to wrestle with monumental massive structures and how they manifest in the weeds, grammars, and habits of FDWP environmental research. Terraformatics is a situating analytic (starting with place-based knots) and way of study (impossible to fully realize) that names FDWP as a terraformation of anxious extractive violence. We undertake it from the place of an FDWP-driven university in an effort to refuse FDWP's terms and logics and to help make room for other already active futures. Instead of environment or planet, which are massive objects, we use the word *terraforming* and think about it as a collective action. Thus when we say terraformation, we are not trying to conjure another Charismatic Mega Concept (CMC). Instead, we are striving to understand our research as participating in a provisional ensemble of material, earthly, living, social, affective, imaginative, machinic, spiritual, political, and epistemic processes. Interconnected processes that are hard to name, and which are scaled at the intimate as well as the massive.

This is to say we mean terraformation *to evoke formation as both process and formed form; and in their perdurance they contour conditions. Terraformed and terraforming terraformations are the constitutively cocomposing conditions of particular worldings.*

Those thirty-five words said exactly what we mean, and it took us five years to write them, but they're also unintelligible. Let's try again: What is terraformatics? Terraformatics is not a complete method or codified process. It is our experiment in trying to study differently, and we offer it as an aspirational mode of engagement that needs further help from you. We want terraformatics to offer support when we stumble or slide into subject-verb-object grammars of thought and speech that incline us to assume or posit through implication a world of separate subjects and objects, or humans and environments. This is not mere theoretical nerdery (though perhaps it is also that!), for those very concepts, actions, and grammars, far from being self-evident in experience, are inherited; they were forged as part of the material colonial terraformation of the One World (scientific reason, terra nullius, White Supremacy). Terraformatics is about refusing these One World grammars when one can, while not avoiding the obligation to confront violence, and at the same time participating in making other, less harmful arrangements.

We recognize that we are not the only ones to speak of terraforming and that, in fact, terraforming talk is central to FDWP modes of reacting to what it characterizes as planetary environmental crisis. In seizing and redeploying this term, we are attempting to work against the contemporary terrapolitical imaginaries of the environment and geoengineering with their One World managerialisms, even as we observe them getting stronger and stronger within the universities where we work. Within those expert imaginaries, facing a superheating global horizon, the term *terraforming* is invoked to name a planetary-scaled engineering fix, a knowing nod to its initiation as a speculative concept in science fiction that detailed the engineering and control of environments on other planets. The term then moved in the 1970s into scientific discussions at NASA exploring the possibilities of planetary colonization (mostly of Mars), naming the process whereby a hostile environment could be altered in order to become suitable explicitly for human life.[1] These early terraforming imaginaries are coded as a planetary property grab, drawing on unreconstructed frontier ideologies.

---

1 For an early NASA consideration of terraforming, see Averner and Macelroy (1976). For an assessment of ideas about terraforming Mars, see Fogg (2011); and for broader considerations of terraforming for habitable worlds, see Oberg (1981) and Beech (2009).

4.1 Daein Ballard, *Mars Transition V*, 2006. Digital illustration of the terraforming of Mars. Wikimedia Commons.

This book offers terraformatics as a more humble analytic, a situated premise, one that seeks not to reproduce FDWP but to provincialize and deflate it; but more importantly, it promotes modes of study animating more habitable conditions at modest scales. It does not rush to the planetary or offer a heroic escape pod to another world. We are not headed to Mars. It is not another CMC. In seeking not to reproduce liberal reason, our understanding of terraformations strives to avoid hailing a knowing subject who studies an inert world and then imposes a single grid of intelligibility. We know we will not escape the grip of FDWP; we will not fully succeed. In this way, the invitation to study terraformations is also an invitation to a set of ethical and political commitments, holding a critical stance to the settler state's claims to sovereignty and empire, and a corresponding resolve to acknowledge the world of many worldings.[2]

Further, to study terraformations is to sign up to the project of co-making less hostile worlds in joined-up work, yet keeping our critical apparatus whirring against the many entrenched projects offering up tempting narratives of salvation, protection, or defense. We acknowledge that invitations to join up in salvation missions abound. We saved you by reengineering your economy, yes? Aren't you happy the dangerous people are mass incarcerated? Feeling protected by those air strikes? Want our company to engineer the atmosphere to cleverly save you from climate change? Relieved that the homeless encampment next to the farmers market has been cleaned up? Always remember that FDWP offers itself up as a harm reduction project that actually supports its key audience of billionaires and White men. So many projects of FDWP claim to be terraforming on behalf of a total human collective (this is a tell), but they are simply committing the rest of us to the fallout. By *joined-up study*, we mean nonsolitary study that is accountable beyond the university to communities, places, and nonhuman beings intent on resisting FDWP. Nonhuman beings are plants and animals but also soil and winds, computers and hammers. All of which is to say, committing as scholars to the contouring of less hostile conditions may be crucial, but it takes care, working within sometimes difficult solidarities.

The invitation to study terraformations is not merely a naming exercise, though it matters profoundly, as Haraway (2016a) teaches, what names name, what stories story. It matters that terra is presently already a way of naming place, land, earth, and worlds. This invitation then is not about finding just any old remaking of material conditions. The terraformatic invitation as we offer it carries particular intellectual and political commitments. Building a less hostile condition binds what goes on in the seminar room with climate—linking the politics of critical thought with ways of being with others even while keeping injurious formations under review. It understands that beings and forces, and not just people, are animate potential allies to be joined up with. This is not easy, because conditions are full of incommensurate needs, contradictions, and injurious entanglements in each other's possibilities for living.

---

2 For considerations of ontologically incommensurable worlds, see de la Cadena (2015); Omura et al. (2018); and Povinelli (2001, 2016).

Terraformations are not things. They are not infrastructures, for example, or technologies, even though that is a part of it. They are not conglomerations of things, such as assemblages or networks. They are processual, patterned, and political, material, social, affective relations that emerge from and then subsequently alter conditions. That is, terraformations are literally terraforming, they are made from conditions and are also conditioning. They are actions in the middle of actions. Verbing all the way down. They are made of terraforming enterprises that come in many sizes and forms. Terraformations can be the project of making less hostile worlds in situated collective conditions; they can also be the injurious colonial projects of surveying land in order to steal it, or a nuclear state looking for a so-called test site to detonate its next-generation weapons. In other words, terraformations are not inherently good. Terraformations can be the project of coupling anti-Black dehumanization with industrial agriculture to make plantation slavery and its wake; it can also be the project of blocking a road to stop a pipeline, which is at the same time an insistence on Indigenous sovereignty, a joined-up sociality, a risk to your life, and an honoring of the livingness of stone and water. They are instantiations of relations in human and more-than-human forms that require careful study to understand in their non-universal specificity. They are in us, all around us, and they can be otherwise. Studying from another incommensurate worlding, you may not want to use the terms of terraformatics at all as part of your processual work. That can make sense too. This provincialized and provisional sense of terraformatics has been helpful to us as we try to engage with projects that change conditions and critically acknowledge our participation in them already.

So what do we really mean when we say *conditions*? We are using this word to foreground the material composition and state of relations of beings, makings, doings, and forces as a potentiating realm. Part of our resolve is that we feel obligated to materiality in its many formings. Conditions are not a static container you are inside; we see them as processual relations often materially instantiated that you are a part of and made through. You shape conditions and are shaped by them, but you can't simply control or manage them. Conditions are populated by phenomena-in-the-making, even if some of this making is very slow and durable. Conditions therefore can be very hard to alter. At the same time, conditions as a term tries not to assume the objects, domains, technologies, forces, and feelings are just there, a priori. Instead, conditions are a nonuniversal, non-object-assuming way to talk about the milieu of relations that one is a part of, that is always a field of coconstituting, in a world of many worlds. In conditions, beings are active in relation to one another, making each other possible, or impossible, composing and dissolving, birthing and destroying, thickening and abandoning.

Here there is a tangle of apprehension that makes linear writing a difficult way to convey terraforming (and hence the baroque nature of so many sentences in this book!). Any apprehension of conditions is built of particularly contoured terraformings, and at the same time conditions include unknown and confounding forces and perturbations. These can be silent disruptions; at other times they are the elephants in the room, the public secrets that are beyond reason. Our apprehension of conditions is at once a recognition of how history and structures are baked into our sensing and knowing apparati, and also alterable by unnoticed deviations and forces. This is in part the parable of Planet Nine and the power of attending to perturbations. Discernment of nonsensical forms and forces may at times illuminate escape hatches, be signals from possible allies, or enable new conditions. It may also be just something you ate.

We use the term *conditions* as a way to respectfully not appropriate Indigenous senses of land, which is not for the taking and should not become a universalized replacement for *environment*. So we in part use *conditions* as an allied prompt to recognize the fulsomeness of relations, confirming our differential obligations to land within colonial relations, without predefining them, without universalizing them, or taking the specificity from each other's worlds. *Conditions* is also a useful open term because it invokes a sense of ambient and substantive material surrounds (such as the weather, the atmosphere, the water, the soil, the mood) and a sense of historical conditions, the genealogies that shape what is possible to say, ask, and do—what Michel Foucault called *conditions of possibility*.[3] Conditions are in the plural. Conditions form and knock us about, whether we know about them or not. As a concept, conditions afford a commitment to combining materialist and political study in a world of many worlds. It involves admitting that we only ever know conditions

---

3 In *The Order of Things*, Foucault ([1966] 2002, xxiv), proposed a mode of examining the historical past "which is not of its growing perfection but of its conditions of possibility; in this account, what should appear are those configurations within the space of knowledge which have given rise to the diverse forms of empirical science."

via particular tools and ways of knowing, often derived from the world-slicing objectifying practices of the university.

A terraformatic study of conditions is thus with and against university modes of research. Study of conditions invites a critical relation to explication, not abandoning it, but juxtaposing it with other modalities of study that defy positivism and management. When doing this, we may find help from conditions themselves, for conditions approached in their fulsomeness easily exceed the reach of reason and One World explication. The crucial proposal of terraformatics as a kind of study for us is how we commit to making less hostile conditions, even at small and intimate scales from the seminar room to the hallway to the published text.

Conditions are different from worlds in this book. Worlds are ontological. They are place-based human modes of being that then are also modes of living. We have not assumed worlds have cultures or natures or politics, as these are domainings that are false universals, and thus are overreaching and epistemically destructive. Worlds can be profoundly incommensurate to one another, and thus to the epistemes of the North American university. If one takes seriously the request not to impose common measure on worlds, to refuse the colonial impulse to bring all into commensurability, to reject the capitalist insistence on fungibility, then worlds crucially may not be any of your business. If you are not part of a world, have not been invited, then you should stay away. It is not for you. Or if you are invited to engage, you are merely a guest. There is no implicit invitation to extract, and in fact, the invitation is to refuse extraction-based research. For some of us, who regularly live in multiple worlds, this is just how it is when you enter the North American university. How, and how not, to bring one's worlds into the university becomes then a question of care and protection. It becomes part of the question of how we are making conditions for ourselves with others, that is, how we terraform across multiple worlds inside universities and outside.

When we say the *world of many worlds*, we mean it. We take seriously the invitation of the Zapatistas to think this way and follow colleagues like Marisol de la Cadena in this approach.[4] The world of many worlds is not just a multiplicity of points of view, it is distinctive and specific contextual modes of being and knowing. The One World of FDWP

---

4   On Zapatista thought, see, Subcomandante Marcos (2022), as well as de la Cadena (2015a) and de la Cadena and Blaser (2018).

seeks to destroy all other worlds. Worldings, rather than world, points us to the speculative and processual project of making modes of being in joined-up composition without appropriating. Worlds are not static or uniform. So when we say a world of many worlds, we are also thinking of the worlds yet to come. The worry that we wrestle with in using the word *world* is that it is burdened by the historical ways anthropologists, geographers, and historians have all too often construed the term through colonial modalities.[5]

Terraformatics, as we envision the practice, is no simple recipe, even as it strives to be humble in its mode of study. It is hard and impossible, absurd and doable all at once. Ultimately, it commits to a project of harm reduction by working consciously to make conditions less violent in joined-up stumbles and steps. What do we mean by *harm reduction*? What distinguishes a less hostile worlding? There is no single answer. The historical achievement of FDWP is to distribute violences—toxic, military, identitarian—and then work to naturalize them as necessary externalities, unfortunate but inevitable events that are the result of the only possible mode of living. FDWP is necropolitical through and through. As an analytic, terraformatics seeks to understand the historical formation of the detailed and second-order conditions of FDWP but also to intervene in their makeup while doing so—to shift the registers, the affects, the

---

5 Anthropology sometimes conflates worlds with cultures or a perspective, and leaves in place a sense of a single underlying reality that then only the expert has proper access to. Historically, the anthropological approach to worlds tends to explicate them, fixing them in habitual genres of static description, once emblematized and immortalized in the olive and gray volumes of the early twentieth-century project to collect descriptions of life and language in the Americas, yet also often persistent in conventions of character, setting, and voice that can ground an ethnographic account. For geography, by comparison, the world is often something to be mapped, layered, navigated, and surveyed. It is something to be assembled, blank spots to fill, the knowledge gaps to cover, uneven development to be explained. This sense of the world is connected to exploration, conquering, and subsumption into property. While much of contemporary geography breaks from some of these earlier world histories, the aim of humanism and universality continues to orient it, even remappings through its theories of theories of its mappings of practice and the particular. The history of science often assumes a One World technological determinism only visible in retrospect that all too easily aligns with empire. Meanwhile, conscripts and converts of earth systems often assume the world is the planet and lead us to geoengineering as the only version of imaginable terraformation. So when we are thinking with you, here in this text, about worldings, we ask you to struggle against these extractive colonial modes.

expectations for how beings can coinhabit land, Earth, and place. We are not building a utopia, but dismantling toward a nonuniversalizable something else.

Terraformatics treats the university as a core space of collision between FDWP and the world of many worldings, fracturing into vast sites of possibility for harm reduction—in the classroom, in libraries, in the alley, and via support for all the underbellies that make up life on and off campus.[6] To acknowledge that we are always already confronted in the North American university by the demands of FDWP is also to say that our departments and fields are often training camps for rehearsing, reinforcing, or absolving hostile propositions—which makes them a superb place to start terraforming otherwise. The possibilities for harm reduction are ever present in the spaces and materials around you, in your institutional relationships and commitments to be sure, but also in your head. Terraforming is working, not just on the object and its undoing but on attending to the seedbed and source, making them so that the forms that emerge from a medium are otherwise.

Terraforming is what happened when we wrote this book, striving to make a place together to work differently yet in obligation, with students, with bees, with chemicals, with breath, with bombs, with Google Docs, and each other.

Impossible, already happening, resolved, and not alone.

4.2 IMPOSSIBLE METHODS FOR TERRAFORMATICS RESEARCH STUDIES

If study is a kind of terraforming, how does this change our practice? We are not sure, but that is okay. We don't expect to have all the answers and have been experimenting in how to ask questions. We do want to offer here some basic tactics that we have used in trying to reach toward another mode of study. These methods are not a step-by-step guide or a recipe for professionalization—they are a bundle of small moves, possibilities to consider and try. Please feel invited to experiment with them or to try something else. These are not instructions for institutional change.

---

6 Important guidance on this perspective comes from Harney and Moten (2013) and la paperson (2017).

They will not diversify your department or fix your discipline. They also will not prevent massive environmental violence or end White Supremacy on their own. Each of these methods is necessarily noninnocent; we could stop and rethink each and every word, or we could exit the university all together. We are not against this. It might even be best. Despite their limits, we found these methods of use as small practices that helped to shift our conditions of study, that moved us to start something else, and helped us to write this book.

Shifting microconditions

Shifting microconditions begins with terraforming your immediate relations. That may sound grand, but there are simple ways to start inhabiting study in this way. One place to start is to trouble where we always start. We have said this before. We find ourselves starting where we are again and again. Maybe start with that. What is the *we* that is your *I*? For the *I* is already messily in conditions: There are forces and matters and ancestors that have made your life; you are already engaged with them, already in a terraforming. It might be the smoke in the air, the wind in your hair, or the fear in your bones. It might be fatigue from the side hustles that keep the roof over your head, or the responsibility toward ancestors and those yet to come. Study is not easy when you are hurting, or when it requires exposure to racist taunts and heterosexist jabs. Let's not kid ourselves—microconditions are not romantic or easy; they are often hard and hard won. Those conditions are more than you; they are the expansive nervous system of our surrounds that are compelling you. You can't get out of all of them.

Microconditions can be challenging to shift because FDWP so strongly arranges the university, and it is wearing to be there. One way to hack this is to reconsider how you gather, and who is gathering. Who and what have made the conditions of your presence here in this way possible? Your collectivity will exceed the university. It must. Your responsibilities do, as do your relations. The microconditions might be at the scale of the facilitated meeting, the street action, the community study group, the dance floor, the bedroom, or the gathering. So we have tended to start with the basics of guesting. Where are you? You are always on land. What do you know about that? This question kicks us off toward fostering the feeling of the other worlds that are here, now, with some possible conspirators. No need to wait for disciplinary validation or institutional permission. This is a different route to study than starting with a problem or an object. Starting with others in microconditions is not about being cozy and nice. Coming together is hard, it is a struggle, it is translation, it won't necessarily work. Even love and care can be appropriated by FDWP. Starting with microconditions is about orienting to a politics of research that is not just about you, the individual, but still right here. Microconditions are not the end game; they are the substrate and a starting place.

No one is alone

Research in an FDWP mode looks for heroes and innovation, mostly White and male, and erases its takings. Citation is political. It is not just a chore. Try getting into it! Who and what needs to be honored, acknowledged, taught, and listened to?[7] What violence, sacrifice, or destruction paid the price? Accept the responsibilities and relations that come with being held up by so many others. All this effort, this being and doing, by others is a source of tremendous strength.

If you are reading this book, then you are already not doing this work alone. Generations of study and action have been working and developing practices, techniques, concepts, and tools, both within and beyond the university, for some time. So ask: What and who are you bringing close? What relations are you building from? Whose words and worlds are you elevating? What itineraries join up in you? Within the terms of these relations, what does it mean to care for this closeness, to acknowledge debts or gifts, and to be studying and acting together? But being with others does not mean you have an entitlement to take from them. Appropriation is the norm in the university. Check in. Do you need to apologize? Do you need to listen? We have had to do many check-ins and apologies with each other in writing this book just because we have been working across differences. Moreover, when we say you are not alone, we don't just mean humans.

---

7   On citation as gratitude, see Liboiron (2021); McKittrick (2021); and Ahmed (2017).

Terraforming for some can be terrabreaking for others

This also means that a project of terraforming, of altering conditions, requires asking what existing formations need to be broken. We have said this in many ways throughout the book, and it is a core commitment. What needs to stop to make more habitable worlds or to dismantle FDWP? Here, we have found it important to remember that making is not a positive and breaking is not a negative. Terraformations are not merely good or bad. Deciding not to reproduce FDWP is vital, but finding how to not subsidize and give energy to the fine details of what has to go is not easy. FDWP wants to make itself live, and builds conditions hospitable for itself. What are you invested in propping up? When one studies, when one is making concepts and arguments, you are terraforming, and thus necessarily involved in reinforcing, escaping, or transforming these violent orders. So think about what needs to go—we call this invoking the sledgehammer—and start swinging. Violence cannot just be met with good intentions. The thing you want to stop might be in your head as much as in your disciplinary conversations or in a practice, or in a worlding you belong in. To break something, to dismantle it, to ignore it, remove support for it, abolish it, has consequences and sometimes is painful. There is usually a hostile counterreaction. In fact it can be dangerous and traumatic.[8] Alter-worldings can have incommensurate needs to break and make. Fortunately we are not the only ones trying to stop FDWP. Phew! Listen for the no's around you. This is not a blank check to destroy (that is the game of FDWP). With abolition and refusal come obligation and opening. So we have tried to use small openings to generate a different practice, to activate a better political method, to think seriously about reducing harms, to listen deeply, and to amplify the multiple worlds that are always present in the room. It is not all about you.

Take one step into something else

Dismantling colonialism, ending patriarchy, defending land is so daunting. Plus, it's not going to succeed anytime soon. These are intergenerational collective tasks facing the powerful defense mechanisms of

---

8   See Frantz Fanon (2007, 86–87) on this point, as well as in his clinical psychiatric studies of colonial violence (2018).

FDWP.[9] Global warming is so big. Whiteness is so muscular. The US Supreme Court's antidemocracy rulings seem so protected. At least that is what we have been taught over and over. But remember, each of us is already and has always been terraforming, reproducing, breaking, and making worlds, even if just by floating along with the status quo. Embrace this responsibility! One thing to try is to figure out one doable step. Just one step into something else. Consider not reproducing something (no more extractive research!) or giving up key terms (not going to use that word ever again!), or write something in your own language, and then step into that space. Just one step. There are lots of different spaces that afford and compel in different ways. The one step can also be found by slowing everything down. Study is made up of millions of little steps, habits of thought and behaviors that you can attend to and tune for a different order of being or community. By slowing down, suddenly the project is filled with all the steps that one might take, and it becomes possible to notice that the worlds you are in are replete with subtle invitations to move in other ways. Stay humble, join up with others, and keep taking that one step, even if it is a stumble. It won't be perfect. But it is so doable, and you can do it again and again. This is the quotidian work of terraforming.

Welcome the contradictions

It makes sense to attempt to find a solution to a problem, except that this solutionism at the university, in the state, inside the corporation is a key part of the problem. As we have tried to show, solutionism is surrounded by strategic unknowing, and it is likely to affirm as most reasonable or feasible the nonconsensual intervention, the technofix that leaves all other orders in place. The weapon scientists always solve their problem with another weapon; the oil companies manage their markets by inventing fracking; the police assuage some people's fear by wielding a weapon against someone else; financial capital solves its boom and bust cycles with a new debt instrument. It is tempting to smooth things over, as that is often what disciplines ask of our analysis and narrative. In writing this book we have struggled against this smoothing. When we

---

9 See Weston (2022) for an assessment of the politics of ecological inheritance in which the dangers to future people are the activities of preceding generations.

smooth over the contradictions, double binds, constitutive tensions, and incommensurabilities, we quickly find our study on the side of erasure.

We are all partially stuck in the mud of the dead White world. It has added its toxic elements to our flesh and altered the air we all breathe. Wherever you are, you are to different degrees with and against its violence, both when it seeks to kill you and when it props you up. Rather than normalizing or erasing the simultaneity, try naming this bind, this contradiction. It can be a profound gift to each other to try to name and avow the constitutive tensions that one is in, to not smooth it over. Living in the binds, working the binds, is a crucial and sometimes cruel kind of knowledge making. We have found it helpful to lift up the contradiction and elevate it in the text, to bring the blockage forward and describe it. Refuse the erasure. For some of us, this makes perfect sense because there is no other choice. If you are constantly living in multiple worlds at once, you may be already skilled at this code switching, multidimensional existence. If you are from the world of One World, a first move is to practice letting go of commensurability as a value. At the same time, it is hard to keep sane when the university is telling you over and over that there is only One World. The disciplines, even at their liberal best, expect you to make equivalence, find common measure, expose actions evenly to the law, find symmetry between the sides. In asking for these modes of study, the university is hailing you to continue its colonial reproduction. The world of many worlds is none of this. It is uneven, incommensurate, knotted, contingent, and contradictory. This is where we want to write from, even if it is a struggle, and we stumble much of the time.

Unknowing objects

Learning from Planet Nine, we have aspired to trust that we don't know, and to seek out things that make no sense, that trouble, or are perturbed or perturbing. Does it keep you up at night? Stay with the thought that returns, the objects that excite, the notions and things that haunt. We have tried to hold the perturbation, lightly, to let it move us before being tempted to place it back into intelligibility; that is, back into the One World that locks it into order, where it can be made comparable or equivalent. Engage objects as a subject: We imagine we ask it questions, and that it then questions us. Why are you interested? What motivates you to care? Why not? What conditions (material, conceptual, historical, affective, personal, spiritual, political) matter and why? The site of concern that one studies is always about more than what it is and where it belongs. How does what it does help you unknow, reimagine, or even break things? How does it undo what is rational, what is admissible? We recommend trying to stay with this unknowing as much as you can. When you resist the impulse to return to epistemic coherence and consistency, everything is its own platypus, virus, or parasite. Many relations have the possibility of returning to help unorder the One World. You do not have to put them back, but to be in alliance with them, to combine forces to make worlds and unmake the One World that has conscripted for far too long. So don't try to make everything commensurate or map it onto a single plane. You do not have to be symmetrical, you do not have to conform to the standard measure, and you certainly do not have to reinforce them. Be open to worlds of actions amid actions, verbs turning into more verbs. If you find yourself back on the bummer train, remember you are already intergalactic and can move toward alter-wise relations all around.

The future is in the past; time is part of the work

One of the principal aspirations of One World building is making time and space universal.[10] When we dwell in imperial temporality, then many hopes, many people, many projects, many practices, many timelines, appear to have been destroyed, left behind, or made unfeasible. They are rendered as past, such that history becomes a done deal and just a matter of facts. Imperial time imagines a linear chain of causality that adds up to the One World of FDWP. This imperial time trick can be refused. Step somewhere else, join someone else, be moved, orient to another timeline—brush history against or start somewhere altogether different. There is not just one genealogy; there are many genealogies. These many genealogies are, moreover, not commensurate with one another. They don't measure out different degrees of responsibility for the present condition. They are instead the genealogies of the world of many worlds. They are embattled. They are still active. The past is not over. There are many futures as well. Though imperial history asserts that what is done is done, and that only a singular future holds possibility, the past is in fact thick with other worlds, resistances, wisdoms, and inventions. Peo-

---

10   On time, see Ogle (2015); on mapping, Rankin (2018).

ple have been at this for a long time. The past is full of ancestors or archives or technofossils to guide you, full of past terraforming knots for you to pick up the thread on. The past is active in the present and is part of the terraforming. In fact, descent is not necessarily the most crucial relation; it is one kind of kinship and temporality among many. The official archives are not the evidence of the truthful world. They are the records of FDWP that need to be read as such.[11] The past is so much bigger. Sometimes imagination is the better truth. Study, even when it is about the past, is speculative. This is not anachronism; it is time work.

Accept impurity and commit to harm reduction

Life inside multigenerational violent orders disallows the possibility of a neutral outside, a nonimplicated position from which to stand.[12] So accept this reality and give up on the idea of the purely innocent or objective position. That doesn't mean giving up accountability and obligation. It means there is no point in aiming for the blameless escape. Study is not going to disappear FDWP. But just because failure and noninnocence are inevitable doesn't mean that you are off the hook. If Whiteness hails you as its avatar, and even if your ancestors are from other places, you are still brought into responsibility for its violence because of where you are here and now. If Whiteness wants to kill you, it might also be recruiting you at the same time by promising you some small shares in its entitlements. And if this is not the case, and Whiteness just wants to wreck you, you are in part made of these scars. And then our coping strategies can entangle us back with violence. To say there is no space or position of innocence is not to despair. It is to refuse the ethical formation of innocence and salvation whose logics have structured so many forms of colonial and racist violence already.[13] Its ethics of aspirational innocence has two additional insidious effects. It infuses fragility into bodies such that they cannot bear the weights of structural complicity, and it forecloses

---

11   As Walter Benjamin (1968, 256–7) puts it: "There is no document of civilization which is not at the same time a document of barbarism"; the task then is "to brush history against the grain." On the necessity of imagination in and beyond history, see also Morrison (1995).
12   On the problem of a nonimplicated stance, see Konsmo and Recollet (2019); Shotwell (2016); Ticktin (2017); and Hong (2020).
13   On White innocence, see Wekker (2016) and Ticktin (2017).

tactics of harm reduction. We can learn to recognize and refuse this trap by owning our impurities. There are so many ways of being impure, being a mess and messing up. So many temptations to cover it over. That is why you will make mistakes and need others.

Conspiracy is not always wrong; consider making a conspiracy

Maintaining FDWP requires constant manipulation, gaslighting, and propaganda, efforts to make this singular form seem like it is both immortal and unchangeable and totalizing. Disinformation, lying, intimidation are key tools for maintaining a hyperviolent political order. So don't be afraid to identify the disinformation. The conspiracy of Whiteness is real and ever-present—so watch both your back and your news feed! Also, don't take all this on by yourself; build your own community, gather folks that you can study with in solidarity, search out and find those people you can conspire with—as lungs teach us, we need solidarity in breathing, in conspiring with others, despite the deceptive fogs, the tear gas, the attentional hacks and strategic misdirections. But know that joined-up study is not easy to pull off. Sometimes it's best to take study off world, which is not to Mars, just off the campus of the One World. We often need to protect ourselves from FDWP intensities. The university shouldn't get all of you. Consider being conspiratorial and embracing modes of fugitivity beyond the university.

Welcome no

Refusals and escapes come with withdrawal symptoms. Refusing extractive epistemic practices, White disciplinary norms, and stances of mastery is not trivial, as these formations both surround and saturate us. They are trained into academic infrastructures and disciplined minds. As you step out of their force fields, it is easy to feel fear, loss, disorientation. It may hurt. You may relapse in search of relief. You may desire something that hurts you. That's okay—keep at it. It will get easier. Orienting with others can help. Importantly, refusals are not just about what you say no to. Protecting yourself from the extractions of academia is important. But so too is realizing that consent can be double edged. Conventional research ethics is all about how to get to yes. We are constantly invited to try to choreograph and listen for the yes. But if one is trying to really care about refusal, this requires not only cultivating the art of saying no but also listening for, inviting, and appreciating the nos of others.

Do what you are doing, maybe

You may already be doing this stuff, but maybe not in the university. What have you been doing? With whom, with what? You can build on the work you, and your communities, have done and the lives you have lived for practices, models, and support. What things and words have mattered to you in that life you have been living? Are you already a community organizer? A Land defender? Part of a mutual aid collective? In what ways have you already been engaging with questions about conditions for living, conditions for change, and changing conditions? Shield yourself from the colonial fantasies of the university, prioritize recovery strategies when exposed to toxicities, and seek friends, allies, and accomplices as a way of surviving and working to shift the conditions.

You may not already be doing this stuff. We're all late to the party in some way. Own what draws you here; an awkward honest introduction is better than an elevator pitch. Listen more, talk less. Learn to follow and help out, and you will learn to move.

Mistakes will be made

My fly is down. We are not naming names, but at least one of us has had the experience of looking down after giving a lecture and seeing that their zipper is undone. And maybe this has happened more than once. Okay, maybe stuff like this happens a lot. The missteps, mistakes, and fumbles are all the way along. Not just small, but big. Why wouldn't they be? The doc is filled with typos. I am pretty sure I just mispronounced this word. I'm not just late to the meeting but in the wrong one. We are bumbling through contradictory, incommensurate conditions; we are dodging violence and multitasking all at once, even with our best efforts. It can be very anxious making. It can be hard to sleep. I am not feeling well. It is tempting to try to cover this up. The university is full of performances of mastery. And yet, mastery is a big part of the FDWP problem. Since we are constitutively always going to be stumbling and failing, try owning the misstep. Knowing how to give a good apology is a powerful skill for terraforming microconditions.[14] At first it feels embarrassing, but then it is a relief. Describing how you don't know, are confused, feel lost is part of the study. The first draft is always a hot mess. Try not to hide it. This doesn't mean that anything goes. It means that even when you are doing your very best to be in good relations, there is no way you are not going to mess up. In fact, from the perspective of FDWP, you are going to be in the wrong, be in trouble, or even be accused of being a monster. Taking on FDWP is a long-term, intergenerational project. Don't defer just because you will fail. Failing is constitutive. You don't need to try to fail—it's just going to happen.

Pay attention to desire

In the name of being progressive, your discipline will likely want to train you in negative critique, and so we get very good at tearing each other to shreds. We know we have been trained to be able to pull out our sharp claws. And then, your discipline likely invites you to position yourself as a hero by doing some deficit research, that is, enumerating and excavating the pain and trauma of other beings as a way to respond to violent orders. We hope we have by now made the case in this book that these are modalities of research you can say no to. So what to try instead? Try paying attention to desire. This is something Audre Lorde and Eve Tuck point us to, but also a basic tenet of queer studies.[15] You can start with your own, but that is insufficient. Keep in mind that desire is wily. We can want things that are violent and selfish. Capitalism, colonialism, and Whiteness are filled with desire machines set on luring you in—on telling you what you can and should desire. That is why we say, pay attention to your desire rather than just follow it. Crucially, desire is not an individual emanation of the subject; it circulates. Since you are in a world of many worlds, desire has many lines and forms. Desires join up. And sometimes not. Since you are not alone, desire needs to be at the resolution and resolve of joined-up being. Unevenly, incommensurably, but sometimes resonantly. We are at the edges of each other's desires. This means that we don't have to desire the exact same thing, or appropriate the aspirations

---

14   See Mia Mingus's (2019) writings, talks, and worksheets on apology as an element of proactive accountability.

15   On attending seriously to desire, see Lorde (2007); Tuck (2009); Sedgwick (2008); Muñoz (2009); McClintock (1995); and Harjo (2019).

of someone else, to do joined-up desire-based work. Trying desire-based research is not about just doing whatever excites you. (Though we often get quite excited at times!) But it is about orienting toward the goal of making better relations, better worlds, and hence the ongoing project of terraforming. Desire-oriented research doesn't stop with the refusal, the dismantlement, the breaking. Its ultimate aim is to make more room for, and move into, something else. It is generative. And it can feel pretty darn good.

Free causality from its metaphysical foundation and your study will follow!

The universalized ascription of causality is so baked into scientific and social research that everything is an object linked by cause: both the past and the future, the chick and the egg.[16] So much is at stake. Here we are discussing a certain mode of assigning cause, a version of cause that comes from a mechanist clockwork universe where the researcher sets out to find the deciding factor that can be turned into the lever of governance and control. Who/what causes tropical storms, global pandemics? Does carbon dioxide cause climate change? What/who is responsible for an oil spill or financial collapse? The relations are not necessarily direct, even if the responsibilities run deep. This is why in part we are obsessed with second- and third-order relations and expressions of White Supremacy. What counts as an agent, who makes history, who is held to account, and what is exonerated? What to do when these philosophical traps and epistemic cul-de-sacs litter our path? This is why we have turned to an ongoing querying of how beings and doings are conditioned and conditioning, asking how they become and unravel in terraformations. This is not to say that we abandon cause absolutely, but rather infuse it with ethics, responsibilities, specificities, and contingencies that we are not above. This is why we lean on the word *relation*. Relations, moreover, do not just cause, they influence, obligate, attach, erase, undo, create kin, and so on. Thus, in studying the violences that traffic under the term *environment* we have struggled against mechanistic causal thinking as part

---

16 With apologies for our section heading to George Clinton, and all of Funkadelic, for riffing shamelessly on their 1970 psychedelic soul album *Free Your Mind... and Your Ass Will Follow*.

of FDWP, while holding onto the query of relation and perturbations as ethical bind.

Attend to matterings, not One World materiality

What makes up the terraforming surround of your study? It can be tempting to populate it with objects and materialities that come from earth systems, biology, geology, or from political economy or sociology or science and technology studies, really any discipline's sense of isolating objects and causes. In undertaking anti-FDWP study in a world of many worlds, the beings and doings that surround us refuse to be pinpointed and taxonomized. Our commitment against One World materialism is why we keep starting with a disassembling and multiplying practice of asking again and again, What is an X? In the world of many worlds, an understanding that derives in part from Zapatista Indigenous thought, the invitation is to be humble toward other worldings, to know that other modalities of being and doing have their own matterings that abound, and that there is also so much we do not know, and even other worldings yet to become within. While FDWP wants you to feel like your study is surrounded by it and its materials, we want to provincialize FDWP and remember that this invitation to terraformatics is about humbly orienting to the surround of these abundant conditions and relations in the plural. It is in this abundance that another kind of mattering comes, another kind of terraforming study moved by being and doings that are consequential. So keep orienting toward conditions and their matterings; they make possible what study can become.

Still not sure how to start? Start with middles

What made it possible for you to sit at the table today? Whatever the answer, those conditions of possibility are historical, infrastructural, social, biographical, material, technological, affective, biological, astrological, cultural, chemical, and more. In fact, you might not be able to finish this book because you are already late for work in a job that is harming you but that you cannot refuse because if you do, you or someone you love might suffer. You might not be able to change the space you are in right now because the conditions are not ones that you can shift without more allies, comrades, and some money for rent or child care. You also might not be able to change conditions because there is a virus or a black mold microfungus that is in not only your walls but your lungs.[17] Not all microconditions are subjectable to our immediate alteration. They do not bend to the choosing of the liberal subject. The hard and scary part of rearranging and remaking the order of things in your world is that we often rely on the very conditions that are killing us—and that those conditions are not amenable to management.

This is why we keep trying to answer this question, with as much specificity as possible: What are you in the middle of? What kin helped you get here? Who died? What blockages or traumas inspired the reaction formation that lights your fire? The truth is that we are stuck relying in part on the conditions that are killing us, and we may rightly castigate ourselves for our choices even when choices are so heavily conditioned by the terraformations that we live within. Sometimes even what compels our deepest desires might not just be our aspiration but the counterformation from what we have most suffered. It is a good time to reiterate, there is no terraforming on your own. Being in the middle is a dizzying vortex of multiscalar relations, parallel worlds, uneven forces, and looping times. We have tried not to repress these relations but to find techniques for surviving, riding, and refusing them. We try to soften into them, be moved by them, be surprised and changed by them, protect ourselves from them, be responsible to them. The middle—or better, this middle—is where we started. Yeah, it's a lot. And not nearly enough. The good news is we are all already here!

---

17   As Marx and Engels ([1932] 1998, 42) write, "The phantoms formed in the brains of men are also, necessarily, sublimates of their material life process."

PART 5
CONCLUSION AND FUTURE ASSESSMENT

5.1 WELCOME TO THE END

Not the end times, but something else. At first, it might feel a bit quiet here, a bit of a letdown; the train is long gone, and there are now scant remains of the tracks. If you are still here with us after that rather wild ride, all those undoings and reformings, you might feel like us, that maybe you don't want this to end. The train might be gone but the bummer, alas, is still here. Ouch! You might wonder, as we do, what exactly have we come to, and where does this exactly leave us? But before answering that, it is worth noting a few things. There is no blueprint here at the end; we surely don't offer any prescriptions, and we never had a map or a plan for you. It is certainly not about training—that would root us right back on the same tracks of discipline and mastery. What we have tried to build and stumbled to find is a more grounded and grounding form of collective study. In our efforts to reimagine studying as terraforming, we know we have been in good company, not just the many scholars whose work has shaped us (Stefano Harney and Fred Moten, Linda Tuhiwai Smith, la paperson, Robin D. G. Kelley, Eve Tuck, Sylvia Wynter, to name but a few), but also in the company of manifold conditions beyond the human.

In leaving environment behind, terraforming study orients us to how we become and know in less cramped, more abundant surrounds. The conditions and worldings make study; we do not just study conditions. We have resolved to commit to anti-FDWP practices of terraforming, which for us includes varied responsibilities toward Indigenous worldings and Land Back in North America. Terraforming study invites a commitment to habitability and to making less violent conditions. It also means accepting that we are already humbly porous and altered within the conditions and worlds we rely on, and likely many more than we can fathom.

We have worked toward something else without prescription. We find ourselves moving toward a different kind of spiraling and holding sensation, more grounded and curious. We don't need to all head in the same direction, just step into something else.

5.2 GLEANING GROUP III.5, WORK LOG 21220401, TAMALPAIS ARCHIPELAGO, RSVTERRA9

Diver Mur brought in a curious object found during their weekly rounds at the base of the Bridge Ruins. It was discovered in a wheeled crate lodged under the seat of one of the sunken transport containers. A particularly successful gleaner, Mur always brings back some excellent wire and old tech. But the wheeled crate held a treasure—some hardback books, some faded newspapers, a set of menus from long-gone restaurants serving food we've never heard of, and a single, very odd, paper manuscript.

Titled "Fear of a Dead White Planet," the manuscript was the subject of this week's study night and was read aloud to our collective by our youngest group leader, part of our ongoing project to understand the people who lived here before the floods and the fallouts. The manuscript dates to the era of early experimentation in communal projects right before the collapse, a moment when the consequences of the industrial age were amplifying but not yet fierce enough to force a reorganization of communal life. It took the nuclear exchanges to do that.

We are grateful for the opportunity to engage this odd technofossil, written in an era when the sky was mostly blue and sometimes orange. According to comparative analysis conducted through the Shard Share, the manuscript was likely authored by a group of disaffected middle managers and content producers in the late-subsumption university. Says our Shard Keep:

> *The destiny of the cognitively entangled collective animating this manuscript remains veiled in ambiguity, with speculative inferences pointing toward a coerced diaspora during cataclysmic inundations. An illuminating narrative from a maritime wayfarer in close proximity to the archaeological find hinted at a potential amalgamation with an oxidate syndicalist commune on Pescadero Island. Subsequent generations purportedly evolved into the venerable artisans known for their mastery in kelp alcohol distillation and plastic reconfiguration within the Lisjan communities.*

If the Shard Keep is correct, this places the newspapers in the era organized by financialized money, when militaries protected oil supplies rather than land, water, food, or shelter, and when poison dumping was the norm. More specifically, we suspect the manuscript was prepared at least in part during the COVID Pandemic I years, thus preceding the mass shutdown of the physical universities.

The manuscript has a pleasing rhythm, syncopated with generous stretches of the self-referencing humanities jargon of its time. It also gave us many laughs—so earnest! Were the authors serious? It is remarkably oscillatory in tone—at times aggressive and repetitive, at others inviting and eager to accommodate any imagined reader. The intensities of their struggle to think in their moment about what is now our present is both sobering and strange. As far as we know, none of their oligarchs actually made it to Mars, but those negative-men did launch the geoengineering wars that still affect our world, forcing foundry weeks while we ride out the storms.

Yet what we found most striking in the text were its clear undertones of desperation to think thoughts, tell histories, and make shapes other than those of the Monolith and its Universings. There are also many rough-edged calls for company. We hope they found it—they seem really strung out and hungry for something different. Our Shard Keep offered that many expressive gleanings that come from its era, made in many locations and recorded in a variety of materials, sounded similar tones. But only a handful, at least among the hundreds that circulate in the Share pulse gathering, call for study inside what became the Knowledge Machines. Most of them elected to plant in more promising waters.

It is ironic that the manuscript is so unsatisfied with university practices at precisely the moment that universities began to shed their physical campuses. When Zeta finally bought out most universities, the conversion of students to users to reservoirs for surveillance and information control was complete. After the first malware pandemic erased much of the Zeta cloud, the geoengineering wars burned and then flooded their combustion conversion grids, making our access to the legacy library fragmentary at best. Our past must now be something like their future—bewildering in so many ways.

We used our discussion meal to compare reactions to the manuscript's insistent treatments of study. Pan, of the Condenser Group, mused that the Knowledge Machines may have seemed more ubiquitous, and opportunities for collective land study more scarce, in that age than we ever imagined. It is hard to imagine. Today, in the aftermath of the Geoengineering Rush and the subsequent Weather Wars, most of what we do, from cloud tending and agrosocioarchitecture to genscompositional seashedding, entails communal study. What is the purpose of life activity if not to make more livable conditions?

In agreement, the evening fogs spread themselves across the treetops above our platform. We adjourned to join them.

ACKNOWLEDGMENTS / WORK HISTORY

The relations that have made this odd little book possible are so deep and varied, are formed with and against so much, and are indebted to so many over such a long time that tracing all of its roots is long past being fully possible. There are students, scholars, institutions, fellowships, reviewers, friends, lovers, families, murmurations, and so much more. However, given the themes of this book, it is best to start with the conditions that have brought the book into being.

In many ways, what started this project was a shared collective sense that we were living in the aftermath, and that the mode of understanding and engagement that we each individually had was inadequate and too solitary to go on our own. A sense, shared with many others, that ends were not coming and were instead what we have been struggling with all along. What, in one way or another, brought us to each other was a recognition that we were all in different ways struggling with the scale, scope, and new means of making the world unlivable for so many. We started by building some small modest conditions, finding places with light, places to walk, places to read together or teach, and places to make food. What this collective became for us was a space and sociality that was uncompromising, challenging, and full of love. Even when we could not understand or agree with one another, we would trust that even if one of us might take a different path, the direction, mode of traveling, and company were not only shared but also became the point. The work and the collective were forged together through still wider conditions, dark election cycles, the ends of the Occupy movement, #MeToo, Black Lives Matter, Idle No More, and the genocide in Gaza, all of which are written into the book in ways that pull and orient its understandings, anxieties, and concerns. At times, working and writing as a collective felt like riding a bike without holding the handlebars, and at others, working together felt like coming home. What we knew but learned again and again was that we needed each other, but what became quickly apparent was that we needed and depended on so many, many more.

We started without a goal other than to bring together brilliant students, our talented and sometimes cantankerous colleagues, and amazing longtime comrades to try to make sense of and make the type of solidarity that helps confront powers that deprive us of basic conditions of livability, a collective that made breathing possible and other futures forgeable. We needed to find and create ways and spaces to breathe in, and trying to make this possible, it became increasingly apparent that breathing is a collective endeavor. That meant acknowledging we exist, are free, and breathe only to the degree to which we are bound up with others. It meant acknowledging how porous we are as individuals, that we exist within worlds, and that the only way out of this mess will require ongoing generous and collective solidarity. Any attempt to create new modalities in the university requires persistent work.

We express our deep gratitude to the Neubauer Collegium for Culture and Society at the University of Chicago, which funded the major stages of this project. The unique mission and resources at the Neubauer, including an encouragement to experiment with the form of faculty research and to break disciplinary conventions, were only matched by the extraordinary project support we enjoyed across event structures. We thank the current and former leadership of the Neubauer, Tara Zahra, Jonathan Lear, David Nirenberg, and Elsbeth Carruthers, for their multiple forms of support. And many thanks go to the event and project management team that has made the logistics of this project (navigating complex travel arrangements, leave schedules, and COVID-19 interruptions, to say the least!) not just possible but easy and enjoyable over the years: Mark Sorkin, Carolyn Ownbey, Kristi Bain, Josh Beck, Jamie Bender, Sarah Davies Breen, Bridget Balcom, Annet Lepique, Madeline McKiddy, and Lauren Pachenco. Special thanks also to Julie Marie Lemon, director of the Arts, Science and Culture Initiative at the University of Chicago, for cosponsorship of a series of Chicago events on environmental visualization. These included public presentations by the monumental landscape photographer Michael Light, the data visualization pioneer Laura Kurgan, and the filmmaker Jumana Manna, who brought her remarkable film *Wild Relatives* to Chicago just before COVID shut everything down.

In the spirit of genre-busting collaboration to understand collective conditions, here are some key moments in our coming together. Between conferences, courses, and writing workshops, we also spent regular time on various platforms—Google, Zoom, Piazza, and Slack—as ways of keeping a conversation alive despite all the challenges of working in different cities and navigating the challenges of everyday life and the profound violence of North American politics across war, policing, economy, and environment. One key lesson of the Engineered Worlds (EW) experiment is that the online tools for collaboration allow an easy overcoming of time and space. It is possible now to cowrite as well as to merge seminars across four universities in real time. We hope that by detailing some of our experiments here, we offer others encouragement for their own collaborative experiments inside, across, and beyond disciplinary frames and academy conventions.

Our initial project statement began with a recognition that planetary-scaled problems do not fit within existing disciplines, suggesting that experiments in collaborative multidisciplinary work on environmental conditions (of the kind common in the natural sciences) might be imperative to the social sciences in the twenty-first century. The goal was to generate new theory and concepts recognizing the difficulty and stakes of planetary thinking, and to find ways of connecting very local forms of knowledge to problems of massive scale. Each EW activity was then an experiment in thinking across disciplines, institutions, and generations and an invitation for re-asking our core questions. This produced some new terms but also radically restaged how we were working, with a focus on the status of the North American university as a settler project emerging via the seminars, conferences, and workshops. Together with these concentrated larger events listed below, this strange manuscript was written, rewritten, and rewritten. Its tone and organization is an index of the conditions of its production and the problem space of the university itself.

SPRING 2015—SEMINAR, "ENGINEERED WORLDS"

The coteaching and collective learning started with all our involvement in a seminar at the University of Chicago in the spring of 2015 titled "Engineered Worlds", which Joe organized, inviting Jake, Murphy, and Tim to present work on the problem of the environment. The seminar was devoted to thinking across disciplines in an effort to imagine what a retooled technoscience might contribute to understanding problem sets that link the hyperlocal to the planetary and that have complicated temporal structures. Joe's syllabus was an investigation of scaling problems within environmental science, with a focus on toxic aftermaths. Importantly, Jake, Murphy, and Tim were invited to present work in progress and to share the ways that they individually have come to approach problems of the environment. The results were revelatory, showing how unlearning key concepts was as important as tuning in to specific problems and datasets. The focus on being in the middle of a conundrum, and working to establish a method for staying in the middle, while arguing for more habitable worlds was emergent in this seminar. The seminar was also a means of planning the first conference, which brought together folks we were reading, close friends, and people we had never met but always wanted to engage with. It involved people who helped us think differently, and people we wanted to think and organize with.

Many thanks to the first EW seminar participants, who got us started on such an exceptionally productive note: Damien Bright, Hannah Burnett, Silas Grant, Cameron Hu, Mallory James, Becca Journey, Hiroko Kumaki, Michal Ran-Rubin, Sophia Rhee, Kristen Simmons, Haeden Stewart, and Lauren Sutherland.

OCTOBER 2015—CONFERENCE, "ENGINEERED WORLDS I"

Our first conference, held in Chicago at the Neubauer Collegium, was an interdisciplinary multiday conversation devoted to thinking about the industrial remaking of the world. Beginning with a broad understanding of toxicity, the conference considered the problems of scale, temporality, and living in a biosphere saturated with industrial signatures, across carbon, plastics, and synthetic chemical and nuclear economies. We asked: What new modes of life and altered futures were emerging across species, ecosystems, and atmospheres? What are the modes of social theory and visualization techniques needed to evaluate planetary-scale ecological changes that are both widely distributed and hyperconcentrated within specific populations and localities?

We are grateful to the following participants for their superb papers and collective engagement: Alex Blanchette, David Bond, Deborah Cowen, Elaine Gan, Elizabeth Johnson, Greg Mitman, Nick Shapiro, Beth Povinelli, Nicole Starosielski, and Brett Walker.

Inspired by these exchanges and hungry for more, we decided to try the impossible: to make durable conditions for reading and thinking together, contours for a shared space-time pocket to gather in, even while also in the tracks and rhythms of our separate universities, time zones, home lives, and states. We wished to take up the question of how to study, historicize, and theorize life forms that have been materially altered by the entangled histories of capitalism, militarism, racism, and colonialism, what Murphy calls *alterlife*, and Choy and Zee called *conditions*. And we wanted to do so with students. We wanted to gather a potential intellectual community both for ourselves and for students not contained by departments, subfields, and universities.

So began the first Quadseminar, a four-campus seminar that officially did not exist, yet nonetheless gathered over fifty participants in winter 2016 in seminar rooms on four different campuses linked weekly by synchronous video conference, shared reading notes and discussion threads online, and in-person seminar room discussions that followed quad-campus video meetings. We turned a collective reading list developed after one of our first meetings into a shared syllabus, choosing readings to constellate modes of analysis and specificities of commitment that we knew we wanted to hold together and maintain, even if we did not necessarily know how. Most participants came from graduate programs in science and technology studies, anthropology, cultural studies, geography, history, or women and gender studies, linking the University of Toronto, UC Davis, UC Berkeley, and the University of Chicago. It also included several postdoctoral scholars, auditors, and other visiting students. Together, we pursued the following questions: What forms of alterlife now inhabit our planet? How can we name and study the conditions of emergence and possibility of postnature life from atmospheres to infrastructures? How might we theorize and act for futures beginning from positions of living in the (ongoing) aftermath? What might it mean to embrace emergent forms of alterlife as kin and allies in political struggles and alternative socialities? Our aims were to explore a set of concepts, critical theories, and methods that might grapple with the fallout of radically altered environmental conditions and to compose critical theoretical itineraries and genealogies that might anchor rethinking of the political, conceptual, and historical frameworks for understanding life and its conditions. Bolstered by the generosities of department staff who helped us synchronize our teaching times, as well as countless moments of can-do and goodwill from participants, we collectively muddled through the affordances and recalcitrances of Canvas, SmartSite, Piazza, Skype, Zoom, Slack, and Google Docs, laughed at uneven hardware performance, rolled with mixed time zone and university calendar tempos; and we learned (and learned to learn) from the frictions and gaps that sometimes appeared between different disciplinary norms among our students and ourselves.

We are very grateful to the following seminar participants for leaning into the experimental nature of the seminar and for their serious engagement. At Berkeley, Alexander Arroyo, V. Black, Alex Blanchette, Phillip Campanile, Seth Denizen, Lindsey Dillon, Rebecca Farmer, Seamus Land, Matt Libassi, Jamie Malot Chantry, Juan Marquis Knight, Melina Packer, Eric Peterson, Mary Shi, Julia Sizek, Stathis Yeros, Leonora Zoninsein, and Jerry Zee. At Chicago, Emily Bock, Claire Bowman, Damien Bright, Hannah Burnett, Molly Cunningham, Silas Grant, Lashaya Howie, Marc Kelley, Sophie Reichert, Steven Schwartz, Lauren Sutherland, Kelsye Turner, Kaya Williams, and Yukun Zeng. At Davis, Benjamin Blackman, Tory Brykalski, Ari Conterato, Robert Deakin, Lindsay Dillon, Mariel Garcia Llorens, Marie McDonald, Anne O'Connor, Kristi Onzik, Mel Salm, Sofia Rivera Sotelo, Joanna Stenhardt, Joshua Weiss, and Juan Camilo Cajigas. At Toronto, Jessica Broe-Vayda, Anita Buragohain, Juliette Dupre, Zeinab Farokhi, Sophia Jaworski, Nada Khalifa, Zixian Liu, Alexandra Maris, Isabelle Maurice-Hammond, Cynthia Morinville, Navreet Nahal, Alexandre Paquet, Parnisha Sarkar, Veronica Yeung, and Martina Schluender.

MAY 2017—CONFERENCE, "ENGINEERED WORLDS II:
ON RESOLUTION AND RESOLVE"

The first Quadseminar led directly to a set of follow-up questions informing a second gathering in Chicago in the spring of 2017. The basic premise that arose directly out of our alterlife seminar and discussions was to consider more deeply how to locate oneself (physically, psychically, politically) within the global environment. The shifting nature of earth systems today rides on specific forms of the industrial economy, on historic practices of dispossession and war, and enduring colonial-settler violence. Thus, people are always already inside a set of conditions and aftermaths, but unequally so. This makes the key question of our age one of attunement: How do individuals both come to understand their historical embeddedness in an industrial condition and imagine alternative futures? How do we collectively create a politics that responds to these ongoing aftermaths? This conference interrogated the optics and affects of self-conscious world-making today in light of the mounting effects of industrial toxicity, financial insecurity, and permanent war. In confronting such complex and interlocking problems, we asked: What kinds and degrees of resolution are needed to both identify and locate oneself within conditions that are simultaneously local and planetary? What forms of political resolve can enable the modification of a life already altered by the interplay of toxicity, capitalism, and militarism? How might we imagine a politics not solely of sociality and institutions but of the making of conditions and worlds that are themselves the grounds for emancipatory collectives? It was from these discussions that the concept of terraformation emerged.

For this event we organized graduate student response teams for each panel, hoping to practice the idea of a scholarly network working across not only disciplines but also generations. Our profound thanks to all the EW II conference presenters: Shannon Cram, Nerea Calvillo, Eric Cazdyn, Stephanie Graeter, Kai Wood Mah, Amy McLachlan, Patrick Lynn Rivers, Kathryn Yusoff, and Jerry Zee, as well as the response team members: Emily Bock, Damien Bright, Hannah Burnett, Ella Butler, Molly Cunningham, Silas Grant, Cameron Hu, Emily Simmonds, Kirstin Simmons, Lauren Sutherland, Sophia Rhee, Kaya Williams, and Hannah Woodroofe.

FALL 2019—QUADSEMINAR 2, "ENGINEERED WORLDS:
TERRAFORMATIONS"

When this second Quadseminar course, "Engineered Worlds: Terraformations," took place in the fall of 2019, we took up the insights of the previous spring in Chicago as we sought to reform *terraformations* as a critical term, one that points beyond the geoengineering imaginary to the always situated and always political processes of geosocial formation in any world-scaping. With the prompt of starting with land, we read and worked together to consider how the most powerful terraforming enterprises in human history have been on planet earth rather than off-planet, enterprises tied to the cumulative force of capitalism and colonialism, linking resource extraction to nuclear nationalism to industrial agriculture to urbanization, exploring just how deeply certain terraforming practices built, rather than ameliorated, hostile worlds. A central question of this seminar concerns how progress has been ideologically constructed via a range of violent practices that now constitute the problems we call global warming, the war on terror, neoliberalism, and environmental toxicity. By attending to the pluriverse of terraformations, we also asked: What other visions and practices of world building are possible? How have core concepts of land, world, earth, and planet been forged in noninnocent formations through historically specific infrastructural, conceptual, material, and social relationships? We took up the history and ongoing material consequences of terraforming with serious purpose, multiplying the forms and possibilities of such work across social conditions and land relations.

We thank all the participants in the second Quadseminar for their mode of serious engagement and for pushing out collective thinking. At Berkeley, Ella Belfer, Ataya Cesspooch, Rusana Cieply, Patrick DeSutter, S. Freeman, Patricia Gomes, Chandra Laborde, Robert Moeller, Francisco Morales, Bernardo Moreno, Lauren Pearson, Maria Pettis, Alexii Sigona, Rebecca Struch, Yesenia Trevino, Alien Vega, Morgan Vickers, and Justin Weinstock. At Chicago, Zachary Arrington, Hadeel Badarni, Blair Bainbridge, Alexis Chavez, Hazal Corak, Nikki Grigg, Rachel Howard, Cameron Hu, Hunter Kennedy, Amy McLachlan, Ashima Mittal, Emma Pask, Sophia Rhee, Andrew Seber, and Shuting Zhuang. At Davis, Onur Arsian, Tory Brykalski, Aaron Chavez, Kyoko Chew, Maya Cruz, Kristin Hogue, Marie McDonald, Ivan Montenegro Perini, Nia Shy, Maya Weeks, and Yuting Yin. At Toronto, Jordan Ali, Slaa Attiah, Binta Bajaha, Catherine Barker, Bronwyn Clement, Lauren Duperron, Liam Fox, Yu-han Huang, Zarah Khan, Khalood Kibria, Lindsay LeBlanc, Xiaoyu Liu, Bree Lohman, Aftab Mirzaei, Natalia Mukhina, Adjei Scott, Nisha Toomey, and Atif Khan.

MAY 2020—CONFERENCE 3, "ENGINEERED WORLDS 3:
TERRAFORMATIONS" (CANCELED DUE TO COVID-19)

We had very much hoped that we would have a chance to collectively deepen and remake the concept of terraformations at the final conference in the series and continue to build out a network of scholars, and we were so so close. Though it was slated to be in May 2020, COVID-19 dashed those plans for "Engineered Worlds III: Terraformations," meant to be the concluding chapter in this multiyear project. Our idea was to bring together some of the people who had been on the train since the beginning, along with a new host of intergenerational scholars that we had been reading and thinking about in the seminar all year. It was meant to build on past events in 2015 and 2017, to mobilize insights into individual capacities to effect life on a planetary scale, and shift them into a discussion of how communities are working to live in and move through such violent conditions—that is, to acknowledge, engage, and imagine alternative terraformations. In many ways, the canceling of the conference spurred deeper reading and conversation of these scholars, and though we very much missed the collective reverie of being together, their work and their brilliance were central to the contours of our thinking during the next year as we finished this book.

For being in and influencing our project conversation despite not being able to meet in person in 2020, we thank Gregg Mitman, Lochlann Jain, Ryan Jobson, Hannah Landeker, Vanessa Agard-Jones, Teresa Montoya, Nicholas Shapiro, Chloe Ahmann, Orit Halpern, Hiʻilei Julia Hobart, Tiffany Jeannette King, Jen Rose Smith, Kristen Bos, Brittany Meche, Marie McDonald, Amy McLachlan, Cameron Hu, Damien Bright, Kristen Simmons, Steven Schwartz, Silas Grant, Hannah Burnett, and Aida Sofia Rivera Sotelo.

MAY 2023—FDWP MANUSCRIPT WORKSHOP

Knowing we had a very strange manuscript derived from the many layers of these collaborations, through which our thinking morphed from our collective learning and reactions to political conditions, we knew we needed some outside readers who could help us get out of our own four-way concentrated conversation and into something more like a book. An honest question was whether or not our text would make sense to anyone else. We turned to a set of brilliant and broad thinkers (some who had been with us all along, others who were new to the project) whom we could trust to give us their honest appraisals. We are grateful for their immensely helpful, honest, and creative comments that led to a major revision of the manuscript. We conducted a day-long manuscript workshop on Zoom and came away with a major plan for revising our document.

For their close readings and extremely thoughtful feedback, we thank Abou Farman, Elaine Gan, Alexis Shotwell, Kaya Williams, and Jerry Zee.

In addition to all the people involved in the extended Engineered Worlds project, we also send our gratitude to friends, kin, and colleagues who supported our experiment in myriad ways.

We jointly extend deep appreciation for the support and guidance of our editor, Ken Wissoker, at Duke University Press. In the publishing world, few people really take the long view of investing, not just in individual projects but in authors over their careers, encouraging authors to take chances and try new things, all the while having their backs, engaging with new ideas as they reinvent themselves and their work. Ken, we feel held, seen, engaged, and are oh so very lucky to have you as our editor.

Two anonymous reviewers arranged by Duke University Press generously got on the Intergalactic Bummer Train and offered the most eloquently and brilliantly insightful written reviews we have ever seen, replete with jokes, sage insights, and meaningful encouragement. Their reviews prompted a substantial reorganization of the book that only made it better, and we are deeply grateful for their kind wisdom. We are also grateful to Livia Tenzer, our project editor, for managing the production process with such clarity, expertise, and good cheer.

We thank UC Berkeley and UC Davis Libraries and the librarians who have been engaged in battles with the worst of the corporate publishing machine and for building the infrastructure for alternatives that are the foundations for their open-source support that made possible the publishing of this work free through online access for all in addition to its paper form. Grants from UC Berkeley, UC Davis, and the University of Chicago enabled this text to be open source.

We thank the Oxidate Collective, which influenced the thinking of many of us and which was a generous space to bring incubating ideas. We especially feel gratitude for the brilliant gifts and insights of Diane Nelson, who died during the writing of this book but who had an enormous impact on our thinking, not only intellectually but also in modeling how to be a politically committed, ethically obligated, serious and playful scholar of violence.

Joe would like to thank his coauthors for accepting the crazy invitation that arrived out of the blue over a decade ago with "Engineered Worlds?" in the email subject line. This collaboration has exceeded any hopes he held then of what might possibly be accomplished and, of course, it probably saved his life. To find the form of study you want to be in is an immense gift—so thank you, Tim, Jake, and Murphy. Joe is also very grateful to the students who have animated seminars, the Engineered Worlds project, as well as the US Locations workshop over the years. Thanks again to the Neubauer Collegium and the Division of the

Social Sciences at the University of Chicago for multiple forms of support. Joe is indebted to Joe Dumit, Stefan Helmreich, Lochlann Jain, Max Liboiron, Andrew Mathews, Deborah Thomas, Anna Tsing, and the very missed Lauren Berlant for conversations important to his thinking about this work. Finally, Joe would like to thank Shawn Michelle Smith, who has been with this project in various ways right from the beginning, encouraging its eccentricities and enjoying its mad process, while offering a brilliant day-to-day illustration of what it means to commit to making better worlds.

Tim's appreciation for his coauthors—Joe, Jake, and Murphy—defies scale. The pulse of our gathering, thinking, teaching, and writing together has carried me through. Tim is also deeply grateful to Marisol de la Cadena, Joe Dumit, Colin Milburn, and Jim Griesemer for their friendship and for exemplifying what it means to think capaciously, navigate institutions with care, and cultivate a scholarly world where ideas and people can flourish. They, along with Finn Brunton, Con Diaz, Natalia Duong, Tim Lenoir, Emily Merchant, and Kayleigh Perkov, continue to shape UC Davis STS as a space of intellectual vitality and collaboration. In Anthropology, Marisol, Joe, Tarek Elhaik, Cristiana Giordano, Jeffrey Kahn, Alan Klima, Fatima Mojaddedi, Suzana Sawyer, James Smith, Smriti Srinivas, and Li Zhang have fostered a department that nurtures critical and creative practice. Beyond these institutional homes, Tim's thinking has been sustained and expanded by many friendships—collegial, intellectual, and otherwise. Heartfelt thanks and appreciation to Vivian Choi, Nerea Calvillo, Mel Chen, Nick D'Avella, Lieba Faier, Stefanie Graeter, Cori Hayden, Michael Hathaway, Rana Jaleel, Miyako Inoue, Stacey Langwick, Dimitris Papadopoulos, María Puig de la Bellacasa, Natasha Myers, Hugh Raffles, Kaushik Sunder Rajan, Nick Shapiro, Shiho Satsuka, Elaine Gan, Anna Tsing, and Jerry Zee. Immense thanks also to Mariel Garcia, Marie McDonald, and Mercedes Villalba for their research assistance and organizational wizardry, and to Maya Cruz, Ileanna Cheladyn, and, again, Mercedes Villalba, for their generosity and ingenuity in helping me bring these ideas into undergraduate classrooms. Zamira Ha has made this undertaking possible in more ways than I can name; thank you, Zamira, for your vision of what this collaboration makes possible and for the ongoing remaking and becoming that we continue to share, always with heart and an otherwise in mind. Finally, to Nova, whose laugh at a twist cracks the world open, and to Jujube, master of eye contact and other affective arts—thank you, let's play.

Jake would like to start by thanking the members of this collective, Joe, Murphy, and Tim, for showing him that while taking flight with others can be intimidating, the force and beauty of a collective in the right evening light can be impossibly brilliant. Thank you. He would also like to thank his colleagues in the Department of Geography at Berkeley for fostering and making a remarkable intellectual and political interdisciplinary space like no other. In particular, for help with thinking related to this project, he would like to thank Sharad Chari, Desiree Fields, Gill Hart, Jovan Lewis, Nathan Sayre, and Brandi Summers. Comrades at Berkeley who make being here amid, within, and beyond the university an intellectual and intimate community that Jake treasures: Cori Hayden, Charles Hirschkind, Laurel Larsen, Tianna Bruno, Gerónimo Barrera de la Torre, Beth Piatote, Leah Raiford and Michael Cohen, Polomi Saha and Arti Sethi, Shalani Satkunanandan and Sanjay Narayan, Christina Zanfagna and Ducan Allard, Paul Fine and Joanna Reed, and Mary Beth Pudup. Thank you to old friends who started Jake on many of the ideas that have become many of his core intellectual commitments: Bruce Braun, James McCarthy, Geoff Mann, Anand Pandian, Nancy Peluso, Hugh Raffles, and especially my long-term and at this point tired mentors and close friends Donald Moore and Michal Watts for their deep wisdom and endless patience. Other intellectual inspirations and muses who continually make Jake think deeper and broader are Joe Dumit, Jackie Orr, the late great Diane Nelson, Jonathan Metzl, Liz Roberts, Miriam Ticktin, Miranda Trimmer, and his many graduate students, who keep him endlessly challenged and deeply humbled. Special thanks to those who read the early part of this manuscript or workshop ideas directly related to this work: Shannon Cram, Seth Denizen, Jen Smith, and Xander Lenc. To a few who seem to do everything possible to keep Jake from getting work done, especially Drew, Debra, Gavi, Bodi, Sonora, and Tiago, I am forever grateful for you leading Ruby and him into an ever-flowing set of beautiful distractions. To his most beloved people, Tim Mueller, Lochlan Jain, and Mara Loveman, for your love, your brilliance, and for being the keepers of my heart and soul. Jake thanks the farm and his dad, who, together, showed him what love as an ongoing collaborative terraforming practice can make possible. Thanks also to the broad, dedicated, and motley intergenerational family, now Siempre Verde's beating heart, for keeping Jake's nails dirty, the veggies sweet, and the farm alive. To his remarkable mother, who believed in him when all evidence and good sense said otherwise, but most of all, with the very best of my love, Jake dedicates his part of this work to his daughter Ruby and her spectacular Magic.

For Murphy, this book would be nothing without the joyful friendship, beautiful camaraderie, and generous learning with the coauthors that developed over these many years. Murphy is deeply humbled by their care, hosting, humor, and genius through the many ups and downs of writing, work, and life. So much was learned about teaching, writing, and living in the process of coming together. In Toronto, Murphy is grateful to the wonderful Technoscience Research Unit members, including Kristen Bos, Vanessa Gray, Beze Gray, Fernanda Yanchapaxi, Reena Shadaan, and many others who materialize other possibilities for being and learning in the university. Murphy is also thankful to the critical chemical and toxicity crew at the TRU (Rohini Patel, Sajdeep Soomal, Sophia Jaworski, Reena Shadaan, and Vanbasten Noronha de Araujo) for their inspirational work rethinking the very grounds of substance in relation to anticolonialism. Gratitude is also due to the intellectual friendship of colleagues in Toronto and beyond, including Eve Tuck, Kai Recollet, Max Liboiron, Jas Rault, Beth Coleman, Sarah Sharma, T. L. Cowan, Tamara Walker, Jackie Orr, Miriam Ticktin, Dani Harris, Patrick Keilty, Nick Shapiro, Natasha Myers, Banu Subramaniam, the greatly missed and inspirational Diane Nelson, and many others. The wonderful Oxidate Collective is forever owed thanks for its many years of support and co-learning. Much of Murphy's work on this project was done in the Bay Area with the support of Jake Kosek, Cori Hayden, Joe Dumit, Lochlan Jain, Tim Choy, Mara Loveman, Julie Livingston, Vincanne Adams, and the late, beloved Adele Clarke, as well as other conspirators. Murphy acknowledges the Social Science and Humanities Research Council and the Canada Research Chair program for their generous financial support and contributions toward this book. There are also countless debts to the wonderful people who kept Murphy aloft during troughs of life while writing this book through conversation, food, advice, phone calls, flowers, dancing, jokes, fires, and sharing of many kinds, especially the incisive and magnificent Judith Taylor. As ever, Murphy gives heartfelt thanks to their extended world of parents, kin, and kids (Mika and Maceo) who supported the interconnections of heart and mind along the way.

With such a long process, we know we have missed some folks and forces here who have played parts small and big. The point all along was to get somewhere else we did not yet know with others. This ride on the Intergalactic Bummer Train toward perturbations instead of planets needed this fulsome and messy collectivity of support and struggle. There is no single version of joined-up study, and so we are humbled to acknowledge the oversights, stumbles, and strange generativities of this one, and hope for the more that might be prompted.

REFERENCES

Adams, Vincanne. 2023. *Glyphosate and the Swirl: An Agroindustrial Chemical on the Move*. Durham, NC: Duke University Press.

Agard-Jones, Vanessa. 2012. "What the Sands Remember." *GLQ: A Journal of Lesbian and Gay Studies* 18 (2–3): 325–46.

Agyeman, Julian, Robert D. Bullard, and Bob Evans, eds. 2012. *Just Sustainabilities: Development in an Unequal World*. London: Earthscan.

Ahmed, Sara. 2017. *Living a Feminist Life*. Durham, NC: Duke University Press.

Ahuja, Neel. 2016. *Bioinsecurities: Disease Interventions, Empire, and the Government of Species*. Durham, NC: Duke University Press.

Alexander, Michael A., Ileana Bladé, Matthew Newman, John R. Lanzante, Ngar-Cheung Lau, and James D. Scott. 2002. "The Atmospheric Bridge: The Influence of ENSO Teleconnections on Air–Sea Interaction over the Global Oceans." *Journal of Climate* 15 (16): 2205–31.

Alley, Richard B. 2014. *The Two-Mile Time Machine: Ice Cores, Abrupt Climate Change, and Our Future*. Princeton, NJ: Princeton University Press.

Almond, Douglas, Zinming Du, and Anna Papp. 2022. "Favourability Towards Natural Gas Relates to Funding Source of University Energy Centres." *Nature Climate Change* 12: 1122–28. https://doi.org/10.1038/s41558-022-01521-3.

American Bird Conservancy. 2015. "Kaua'i 'ō'ō Song." YouTube video, posted May 18. https://www.youtube.com/watch?v=LrcCK2zERdM.

Anderson, Carol. 2016. *White Rage: The Unspoken Truth of Our Racial Divide*. New York: Bloomsbury.

Anderson, Carol. 2018. *One Person, No Vote: How Voter Suppression Is Destroying Our Democracy*. New York: Bloomsbury.

Anidjar, Gil. 2004. "Terror Right" *CR: The New Centennial Review* 4 (4): 35–69.

Arendt, Hannah. 1958. *The Human Condition*. Chicago: University of Chicago Press.

Asad, Talal. 1993. *Genealogies of Religion: Discipline and Reasons of Power in Christianity and Islam*. Baltimore: Johns Hopkins University Press.

Asaka, Ikuko. 2017. *Tropical Freedom: Climate, Settler Colonialism, and Black Exclusion in the Age of Emancipation*. Durham, NC: Duke University Press.

Austin, David. 2013. *Fear of a Black Nation: Race, Sex, and Security in Sixties Montreal*. Toronto: Between the Lines.

Averner, M. M., and R. D. Macelroy, eds. 1976. *On the Habitability of Mars: An Approach to Planetary Ecosynthesis*. Washington, DC: NASA. https://ntrs.nasa.gov/citations/19770005775.

Bacigalupi, Paolo. 2015. *The Water Knife*. New York: Knopf.

Bacon, Francis. 2009. *The New Organon, or True Direction Concerning the Interpretation of Nature*. Gloucestershire: Dodo Press.

Bainbridge, Blair. 2020. "Toward No Place in Particular: Detection, Dispossession and Noise Management in the American Southwest." MA thesis, Department of Anthropology, University of Chicago.

Baldwin, James. 1984. "On Being White . . . and Other Lies." *Essence* 14 (12): 90–92.

Baldwin, James. 2021. *The Fire Next Time*. New York: Random House.

Ballard, J. G. 1962. *The Drowned World*. London: Gollancz.

Ballard, J. G. 1965. *The Drought*. London: Cape.

Ballard, J. G. 1981. *Hello America*. London: Cape.

Ballestero, Andrea. 2019. *A Future History of Water*. Durham, NC: Duke University Press.

Barthes, Roland. 2013. *Mythologies: The Complete Edition, in a New Translation*. New York: Macmillan.

Batygin, Konstantin, Fred C. Adams, Michael E. Brown, and Juliette C. Becker. 2019. "The Planet Nine Hypothesis." *Physics Reports* 805: 1–53.

Batygin, Konstantin, and Michael E. Brown. 2016. "Evidence for a Distant Giant Planet in the Solar System." *Astronomical Journal* 151 (2): 22.

Beckett, Lois. 2020. "Older People Would Rather Die Than Let COVID-19 Harm US Economy—Texas Official." *Guardian*, March 24. https://www.theguardian.com/world/2020/mar/24/older-people-would-rather-die-than-let-covid-19-lockdown-harm-us-economy-texas-official-dan-patrick.

Beech, Martin. 2009. *Terraforming: The Creating of Habitable Worlds*. New York: Springer.

Benias, Petros C., Rebecca G. Wells, Bridget Sackey-Aboagye, et al. 2018. "Structure and Distribution of an Unrecognized Interstitium in Human Tissues." *Scientific Reports* 8 (4947). https://doi.org/10.1038/s41598-018-23062-6.

Benjamin, Ruha. 2016. "Racial Fictions, Biological Facts: Expanding the Sociological Imagination Through Speculative Methods." *Catalyst: Feminism, Theory, Technoscience* 2 (2): 1–28.

Benjamin, Walter. 1968. "Theses on the Philosophy of History." In *Illuminations*, 217–52. New York: Schocken.

Benson, Etienne S. 2020. *Surroundings: A History of Environments and Environmentalisms*. Chicago: University of Chicago Press.

Berlant, Lauren. 2011. *Cruel Optimism*. Durham, NC: Duke University Press.

Bezos, Jeff. 2019. "Blue Origin 2019: For the Benefit of Earth." YouTube video, posted May 10. https://www.youtube.com/watch?v=GQ98hGUe6FM.

Bhandar, Brenna. 2018. *Colonial Lives of Property: Law, Land, and Racial Regimes of Ownership*. Durham, NC: Duke University Press.

Black, Megan. 2018. *The Global Interior: Mineral Frontiers and American Power*. Cambridge, MA: Harvard University Press.

Blanchette, Alex. 2020. *Porkopolis: American Animality, Standardized Life, and the Factory Farm*. Durham, NC: Duke University Press.

Bond, David. 2022. *Negative Ecologies: Fossil Fuels and the Discovery of the Environment*. Berkeley: University of California Press.

Bong Joon Ho, dir. 2013. *Snowpiercer*. Weinstein Company. 126 minutes, feature film.

Bonneuil, Christophe, and Jean-Baptiste Fressoz. 2017. *The Shock of the Anthropocene: The Earth, History and Us*. New York: Verso.

Borrows, John. 2019. *Law's Indigenous Ethics*. Toronto: University of Toronto Press.

Broecker, Felix, and Karin Moelling. 2019. "What Viruses Tell Us About Evolution and Immunity: Beyond Darwin?" *Annals of the New York Academy of Sciences* 1447 (1): 53–68.

Browne, Simone. 2015. *Dark Matters: On the Surveillance of Blackness*. Durham, NC: Duke University Press.

Burnard, Trevor G., and John D. Garrigus. 2016. *The Plantation Machine: Atlantic Capitalism in French Saint-Domingue and British Jamaica*. Philadelphia: University of Pennsylvania Press.

Burr, Christina Ann. 2006. *Canada's Victorian Oil Town: The Transformation of Petrolia from a Resource Town into a Victorian Community*. Montreal: McGill-Queen's University Press.

Burton, Nylan. 2020. "People of Color Experience Climate Grief More Deeply Than White People." *Vice*, May 14. https://www.vice.com/en/article/people -of-colour-experience-climate-grief-more-deeply-than-white-people/.

Busbea, Larry. 2020. *The Responsive Environment: Design, Aesthetics, and the Human in the 1970s*. Minneapolis: University of Minnesota Press.

Caduff, Carlo. 2015. *The Pandemic Perhaps: Dramatic Events in a Public Culture of Danger*. Berkeley: University of California Press.

Caldeira, Ken, Govindasamy Bala, and Long Cao. 2013. "The Science of Geoengineering." *Annual Review of Earth Planet Science* 41: 231–56.

Canguilhem, Georges. 2001. "The Living and Its Milieu." *Grey Room* 3: 7–31.

Carrington, Damian. 2017. "'Extraordinary' Levels of Pollutants Found in 10km Deep Mariana Trench." *Guardian*, February 13. https://www.theguardian .com/environment/2017/feb/13/extraordinary-levels-of-toxic-pollution -found-in-10km-deep-mariana-trench.

Carrington, Damian. 2022a. "Microplastics Found in Human Breast Milk for the First Time." *Guardian*, October 7. https://www.theguardian.com /environment/2022/oct/07/microplastics-human-breast-milk-first-time.

Carrington, Damian. 2022b. "Toxic Air Pollution Particles Found in Lungs and Brains of Unborn Babies." *Guardian*, October 5. https://www.theguardian .com/environment/2022/oct/05/toxic-air-pollution-particles-found-in-lungs -and-brains-of-unborn-babies.

Carter, Paul. 1989. *The Road to Botany Bay: An Exploration of Landscape and History*. Chicago: University of Chicago Press.

Castanha, Tony. 2015. "The Doctrine of Discovery: The Legacy and Continuing Impact of Christian 'Discovery' on American Indian Populations." *American Indian Culture and Research Journal* 39 (3): 41–64.

CDP. 2017. *The Carbon Majors Database: CDP Carbon Majors Report 2017*. London: CDP. https://cdn.cdp.net/cdp-production/cms/reports/documents/000/002/327 /original/Carbon-Majors-Report-2017.pdf?1501833772.

Césaire, Aimé. 2000. *Discourse on Colonialism*. New York: NYU Press.

Chakrabarty, Dipesh. 2009. "The Climate of History: Four Theses." *Critical Inquiry* 35 (2): 197–222.

Chakrabarty, Dipesh. 2021. *The Climate of History in a Planetary Age*. Chicago: University of Chicago Press.

Choy, Timothy. 2011. *Ecologies of Comparison: An Ethnography of Endangerment in Hong Kong*. Durham, NC: Duke University Press.

Choy, Timothy. 2021. "Externality, Breathers, Conspiracy: Forms for Atmospheric Reckoning." In *Reactivating Elements: Chemistry, Ecology, Practice*, edited by Dimitris Papadopoulos, María Puig de la Bellacasa, and Natasha Myers, 231–56. Durham, NC: Duke University Press.

Choy, Timothy, and Jerry Zee. 2015. "Condition—Suspension." *Cultural Anthropology* 30 (2): 210–23.

Climate Accountability Institute. 2019. *Carbon Majors: Update of Top Twenty Companies 1965–2017*. https://3vu.742.myftpupload.com/wp-content/uploads /2020/12/CAI-PressRelease-Top20-Oct19.pdf.

Cole, Alyson, and George Shulman. 2019. *Michael Paul Rogin: Derangement and Liberalism*. New York: Routledge.

Collins, Patricia Hill. 2002. *Black Feminist Thought: Knowledge, Consciousness, and the Politics of Empowerment*. New York: Routledge.

Connolly, William E. 1994. "Tocqueville, Territory and Violence." *Theory, Culture and Society* 11: 19–40.

Connolly, William E. 2019. *Climate Machines, Fascist Drives and Truth*. Durham, NC: Duke University Press.

Cook, Katsi. 1997. "Women Are the First Environment." *Native Americas* 14 (3): 58.

Cooper, Anthony H., Teresa J. Brown, Simon J. Price, et al. 2018. "Humans Are the Most Significant Global Geomorphological Driving Force of the 21st Century." *Anthropocene Review* 5 (3): 222–29.

Corporate Accountability. 2023. *Destruction Is at the Heart of Everything We Do: Chevron's Junk Climate Action Agenda and How It Intensifies Global Harm*. https://corporateaccountability.org/resources/chevrons-junk-agenda-report/.

Coudrain, Anne, Matthieu Le Duff, and Danielle Mitja. 2022. "The Anthropocene Is Shifting the Paradigm of Geosciences and Science." *Comptes Rendus Géoscience* 355 (S1): 1–18.

Coulthard, Glen. 2014. *Red Skin, White Masks: Rejecting the Colonial Politics of Recognition*. Minneapolis: University of Minnesota Press.

Cronon, William. 1995. *Uncommon Ground: Rethinking the Human Place in Nature*. New York: W. W. Norton.

Crutzen, Paul J., and Eugene F. Stoermer. 2000. "The 'Anthropocene.'" *Global Change Newsletter* 41: 17.

Davenport, Coral. 2018. "Trump Administration Unveils Its Plan to Relax Car Pollution Rules." *New York Times*, August 2, 2018. https://www.nytimes.com /2018/08/02/climate/trump-auto-emissions-california.html.

Davis, Angelique M., and Rose Ernst. 2019. "Racial Gaslighting." *Politics, Groups, and Identities* 7 (4): 761–74.

Davis, Heather. 2022. *Plastic Matter*. Durham, NC: Duke University Press.

Davis, Heather, and Zoe Todd. 2017. "On the Importance of a Date, or Decolonizing the Anthropocene." *Acme: An International Journal for Critical Geographies* 16 (4): 761–80.

Davis, Mike. 1990. *City of Quartz: Excavating the Future in Los Angeles*. New York: Verso.

de la Cadena, Marisol. 2015a. *Earth Beings: Ecologies of Practice Across Andean Worlds*. Durham, NC: Duke University Press.

de la Cadena, Marisol. 2015b. "Uncommoning Nature." *E-flux*, no. 65. https:// www.e-flux.com/journal/65/336365/uncommoning-nature/.

de la Cadena, Marisol, and Mario Blaser, eds. 2018. *A World of Many Worlds*. Durham, NC: Duke University Press.

Deloria, Vine. 1988. *Custer Died for Your Sins: An Indian Manifesto*. Norman: University of Oklahoma Press.

Deloria, Vine. 2018. *Red Earth, White Lies: Native Americans and the Myth of Scientific Fact*. Lakewood, CO: Fulcrum.

DeLoughrey, Elizabeth M. 2013. "The Myth of Isolates: Ecosystem Ecologies in the Nuclear Pacific." *Cultural Geographies* 20 (2): 167–84.

Demuth, Bathsheba. 2019. *Floating Coast: An Environmental History of the Bering Strait*. New York: W. W. Norton.

De Vos, Jurriaan M., Lucas N. Joppa, John L. Gittleman, Patrick R. Stephens, and Stuart L. Pimm. 2014. "Estimating the Normal Background Rate of Species Extinction." *Conservation Biology* 29 (2): 452–62.

Di Chiro, Giovanna. 2008. "Living Environmentalisms: Coalition Politics, Social Reproduction, and Environmental Justice." *Environmental Politics* 17 (2): 276–98.

Di Chiro, Giovanna. 2017. "Welcome to the White (M)Anthropocene? A Feminist-Environmentalist Critique." In *Routledge Handbook of Gender and Environment*, edited by Sherilyn MacGregor. New York: Routledge.

Dodds, Klaus, and Jen Rose Smith. 2023. "Against Decline? The Geographies of Temporalities of the Arctic Cryosphere." *Geographical Journal* 189 (3): 388–97. https://doi.org/10.1111/geoj.12481.

Doel, Ronald E. 2003. "Constituting the Postwar Earth Sciences: The Military's Influence on the Environmental Sciences in the USA After 1945." *Social Studies of Science* 33 (5): 635–66.

Doherty Sarah J., Philip J. Rasch, Robert Wood, et al. 2023. "An Open Letter Regarding Research on Reflecting Sunlight to Reduce the Risks of Climate Change." February 27. https://climate-intervention-research-letter.org/.

Dorries, Heather. 2022. "What Is Planning Without Property? Relational Practices of Being and Belonging." *Environment and Planning D: Society and Space* 40 (2): 306–18. https://doi.org/10.1177/02637758211068505.

Drinnon, Richard. 1997. *Facing West: The Metaphysics of Indian-Hating and Empire-Building*. Norman: University of Oklahoma Press.

Dris, Rachid, Johnny Gasperi, Vincent Rocher, et al. 2015. "Microplastic Contamination in an Urban Area: A Case Study in Greater Paris." *Environmental Chemistry* 12 (5): 592–99.

Du Bois, W. E. B. (1903) 2009. *The Souls of Black Folk: A Library of America Paperback Classic*. New York: Library of America.

Du Bois, W. E. B. (1920) 1999. *Darkwater: Voices from Within the Veil*. Mineola, NY: Dover.

Du Bois, W. E. B. (1935) 1998. *Black Reconstruction in America, 1860–1880*. New York: Free Press.

Du Bois, W. E. B. (1940) 1983. *Dusk of Dawn: An Essay Toward an Autobiography of a Race Concept*. New York: Routledge.

Dumit, Joseph. 2014. "Writing the Implosion: Teaching the World One Thing at a Time." *Cultural Anthropology* 29 (2): 344–62.

Dumit, Joseph. 2021. "Substance as Method: Bromine, for Example." In *Reactivating Elements: Chemistry, Ecology, Practice*, edited by Dimitris Papadopoulos, María Puig de la Bellacasa, and Natasha Myers, 84–107. Durham, NC: Duke University Press.

Edwards, Paul N. 1996. *The Closed World: Computers and the Politics of Discourse in Cold War America*. Cambridge, MA: MIT Press.

Edwards, Paul N. 2010. *A Vast Machine: Computer Models, Climate Data, and the Politics of Global Warming*. Cambridge, MA: MIT Press.

Eilperin, Juliet, Brady Dennis, and Chris Mooney. 2018. "Trump Administration Sees a 7-Degree Rise in Global Temperatures by 2100." *Washington Post*, September 28. https://www.washingtonpost.com/national/health-science/trump-administration-sees-a-7-degree-rise-in-global-temperatures-by-2100/2018/09/27/b9c6fada-bb45-11e8-bdc0-90f81cc58c5d_story.html.

Elden, Stuart. 2009. *Terror and Territory*. Minneapolis: University of Minnesota Press.

Eriksson, Mats, Patric Lindahl, Per Roos, Henning Dahlgaard, and Elis Holm. 2008. "U, Pu, and Am Nuclear Signatures of the Thule Hydrogen Bomb Debris." *Environmental Science and Technology* 42 (13): 4717–22.

Escobar, Arturo. 2020. *Pluriversal Politics: The Real and the Possible*. Durham, NC: Duke University Press.

Estes, Nick. 2021. "Bill Gates Is the Biggest Private Owner of Farmland in the United States. Why?" *Guardian*, April 5. https://www.theguardian.com /commentisfree/2021/apr/05/bill-gates-climate-crisis-farmland.

Evans, Brad, and Julian Reid. 2014. *Resilient Life: The Art of Living Dangerously*. Cambridge: Polity.

Eze, Emmanuel Chukwudi, ed. 1997. *Race and the Enlightenment: A Reader*. Oxford: Blackwell.

Fanon, Frantz. 2007. *The Wretched of the Earth*. New York: Grove.

Fanon, Frantz. 2008. *Black Skin, White Masks*. New York: Grove.

Fanon, Frantz. 2018. *Alienation and Freedom*. New York: Bloomsbury.

Farish, Matthew. 2010. *The Contours of America's Cold War*. Minneapolis: University of Minnesota Press.

Farman, Abou. 2017. "Terminality." *Social Text* 35 (2): 93–118.

Farman, Abou. 2020. *On Not Dying: Secular Immortality in the Age of Technoscience*. Minneapolis: University of Minnesota Press.

Feigenbaum, Anna. 2017. *Tear Gas: From the Battlefields of World War I to the Streets of Today*. New York: Verso.

Fleming, James Rodger. 2011. *Fixing the Sky: The Checkered History of Weather and Climate Control*. New York: Columbia University Press.

Fleming, James Rodger. 2017. "Excuse Us, While We Fix the Sky: WEIRD Supermen and Climate Engineering." *RCC Perspectives: Transformations in Environment and Society* 4: 23–28. https://www.environmentandsociety.org/perspectives /2017/4/article/excuse-us-while-we-fix-sky-weird-supermen-and-climate -engineering.

Fogg, Martyn J. 2011. "Terraforming Mars: A Review of Concepts." In *Engineering Earth*, edited by Stanley D. Brunn, 2217–25. Dordrecht: Springer Netherlands.

Fortun, Kim. 2012. "Ethnography in Late Industrialism." *Cultural Anthropology* 27 (3): 446–64.

Foucault, Michel. (1966) 2002. *The Order of Things: An Archaeology of the Human Sciences*. New York: Routledge Classics.

Friedman, Lisa. 2020. "E.P.A., Citing Coronavirus, Drastically Relaxes Rules for Polluters." *New York Times*, March 26. https://www.nytimes.com/2020/03/26 /climate/epa-coronavirus-pollution-rules.html.

Gilio-Whitaker, Dina. 2019. *As Long as Grass Grows: The Indigenous Fight for Environmental Justice from Colonization to Standing Rock*. Boston: Beacon.

Gilmore, Ruth Wilson. 2023. *Change Everything: Racial Capitalism and the Case for Abolition*. Chicago: Haymarket.

Gokee, Cameron, Haeden Stewart, and Jason De Leon. 2020. "Scales of Suffering in the US-Mexico Borderlands." *International Journal of Historical Archaeology* 24: 823–51.

Goldstein, Alyosha, ed. 2014. *Formations of United States Colonialism*. Durham, NC: Duke University Press.

Golley, Frank Benjamin. 1993. *A History of the Ecosystem Concept in Ecology*. New Haven, CT: Yale University Press.

Gómez-Barris, Macarena. 2017. *The Extractive Zone: Social Ecologies and Decolonial Perspectives*. Durham, NC: Duke University Press.

Greenfield, Patrick. 2023. "Revealed: More Than 90% of Rainforest Carbon Offsets by Biggest Provider Are Worthless, Analysis Shows." *Guardian*, January 18. https://www.theguardian.com/environment/2023/jan/18/revealed-forest -carbon-offsets-biggest-provider-worthless-verra-aoe.

Greenpeace. 2022. "Circular Claims Fall Flat Again." October 24. https://www .greenpeace.org/usa/reports/circular-claims-fall-flat-again/.

Günel, Gökçe. 2019. *Spaceship in the Desert: Energy, Climate Change, and Urban Design in Abu Dhabi*. Durham, NC: Duke University Press.

Halpern, Orit. 2015. *Beautiful Data: A History of Vision and Reason Since 1945*. Durham, NC: Duke University Press.

Halpern, Orit, and Gökçe Günel. 2017. "Demoing unto Death: Smart Cities, Environment, and Preemptive Hope." *Fibreculture Journal* 144. https://doi.org /10.15307/fcj.29.215.2017.

Hamblin, Jacob Darwin. 2011. *Oceanographers and the Cold War: Disciples of Marine Science*. Seattle: University of Washington Press.

Hamblin, Jacob Darwin. 2013. *Arming Mother Nature: The Birth of Catastrophic Environmentalism*. Oxford: Oxford University Press.

Hamilton, Clive. 2013. *Earthmasters: The Dawn of the Age of Climate Engineering*. New Haven, CT: Yale University Press.

Hamilton, Clive. 2016. "The Anthropocene as Rupture." *Anthropocene Review* 3 (2): 93–106.

Haraway, Donna. 1988. "Situated Knowledges: The Science Question in Feminism and the Privilege of Partial Perspective." *Feminist Studies* 14 (3): 575–99.

Haraway, Donna. 2015. "Anthropocene, Capitalocene, Plantationocene, Chthulucene: Making Kin." *Environmental Humanities* 6 (1): 159–65.

Haraway, Donna. 2016a. "Sowing Worlds: A Seed Bag for Terraforming with Earth Others." In *Staying with the Trouble: Making Kin in the Chthulucene*, 117–25. Durham, NC: Duke University Press.

Haraway, Donna. 2016b. *Staying with the Trouble: Making Kin in the Chthulucene*. Durham, NC: Duke University Press.

Harjo, Laura. 2019. *Spiral to the Stars: Mvskoke Tools of Futurity*. Tucson: University of Arizona Press.

Harney, Stefano, and Fred Moten. 2013. *The Undercommons: Fugitive Planning and Black Study*. New York: Minor Compositions.

Harris, Cheryl. 1993. "Whiteness as Property." *Harvard Law Review* 106 (8): 1707–91.

Hecht, Gabrielle. 2018. "Interscalar Vehicles for an African Anthropocene: On

Waste, Temporality, and Violence." *Cultural Anthropology* 33 (1): 109–41.
https://doi.org/10.14506/ca33.1.05.

Heinlein, Robert A. 2013. *Man Who Sold the Moon / Orphans of the Sky*. Riverdale,
Canada: Baen.

Helmreich, Stefan, Natasha Myers, Sophia Roosth, Michael Rossi, Katrin Klingan,
and Nick Houde. 2022. *What Is Life?* Leipzig: Spector Books.

High Meadows Environmental Institute. 2024. "History and Mission." Princeton
University. https://environment.princeton.edu/about/.

Hobart, Hiʻilei Julia Kawehipuaakahaopulani. 2022. *Cooling the Tropics: Ice, Indigeneity and Hawaiian Refreshment*. Durham, NC: Duke University Press.

Höhler, Sabine. 2015. *Spaceship Earth in the Environmental Age, 1960–1990*. London:
Routledge.

Holden, Emily. 2020. "Over 5,600 Fossil Fuel Companies Have Taken at Least
$3bn in US COVID-19 Aid." *Guardian*, July 7. https://www.theguardian.com
/environment/2020/jul/07/fossil-fuel-industry-coronavirus-aid-us-analysis.

Hong, Cathy Park. 2020. *Minor Feelings: An Asian American Reckoning*. New York:
One World.

hooks, bell. 1986. "Sisterhood: Political Solidarity Between Women." *Feminist Review* 23 (1): 125–38.

hooks, bell. 1991. "Theory as Liberatory Practice." *Yale Journal of Law and Feminism* 4 (1): 1–12.

hooks, bell. 2007. *Ain't I a Woman: Black Women and Feminism*. Boston: South End.

Hoyle, Fred. 1957. *The Black Cloud*. London: Heinemann.

Hu, Shijian, Janet Sprintall, Cong Guan, et al. 2020. "Deep-Reaching Acceleration
of Global Mean Ocean Circulation over the Past Two Decades." *Science Ad-
vances* 6 (6). https://www.science.org/doi/10.1126/sciadv.aax7727.

Hu, Tung-Hui. 2016. *A Prehistory of the Cloud*. Cambridge, MA: MIT Press.

Hubbard, Tasha. 2014. "Buffalo Genocide in Nineteenth-Century North America:
'Kill, Skin, and Sell.'" In *Colonial Genocide in Indigenous North America*,
edited by Alexander Laban Hinton, Andrew Woolford, and Jeff
Benvenuto. Durham, NC: Duke University Press. https://doi.org/10.1215
/9780822376149-014.

Hudson, P. J. 2019. *Bankers and Empire: How Wall Street Colonized the Caribbean*.
Chicago: University of Chicago Press.

Intergovernmental Panel on Climate Change (IPCC). 2018. *Global Warming of 1.5 °C.
An IPCC Special Report on the Impacts of Global Warming of 1.5 °C above Pre-
Industrial Levels and Related Global Greenhouse Gas Emission Pathways, in the
Context of Strengthening the Global Response to the Threat of Climate Change,
Sustainable Development, and Efforts to Eradicate Poverty*. Cambridge: Cam-
bridge University Press. https://doi.org/10.1017/9781009157940.

Intergovernmental Panel on Climate Change (IPCC). 2001. *Climate Change 2001:
The Scientific Basis*. Cambridge: Cambridge University Press. https://www
.ipcc.ch/site/assets/uploads/2018/03/WGI_TAR_full_report.pdf.

Jackson, Zakiyyah Iman. 2020. *Becoming Human: Matter and Meaning in an Antiblack World*. New York: NYU Press.

Jain, Lochlann. 2013. *Malignant: How Cancer Becomes Us*. Berkeley: University of California Press.

Jameson, Fredric. 1998. "The Antinomies of Postmodernity." In *The Cultural Turn: Selected Writing on the Postmodern, 1983–1998*. New York: Verso.

Jemison, N. K. 2015. *The Fifth Season*. New York: Orbit.

Jemison, N. K. 2016. *The Obelisk Gate*. New York: Orbit.

Jemison, N. K. 2017. *The Stone Sky*. New York: Orbit.

Johnson, Lyndon. 1962. *Remarks at Southwest Texas State University*. Texas Archive of the Moving Image. Video. https://texasarchive.org/2010_00003?b=1490&e=1510.

Judy, Ronald AT. 2006. "Provisional Note on Formations of Planetary Violence." *boundary 2* 33 (3): 141–50.

Kaplan, Amy. 2002. *The Anarchy of Empire in the Making of U.S. Culture*. Cambridge, MA: Harvard University Press.

Kaplan, Sarah, Simon Ducroquet, Bonnie Jo Mount, Frank Hulley-Jones, and Emily Wright. 2023. "Hidden Beneath the Surface." *Washington Post*, June 20. https://www.washingtonpost.com/climate-environment/interactive/2023/anthropocene-geologic-time-crawford-lake/.

Karuka, Manu. 2019. *Empire's Tracks: Indigenous Nations, Chinese Workers, and the Transcontinental Railroad*. Berkeley: University of California Press.

Kaufman, Scott. 2012. *Project Plowshare: The Peaceful Use of Nuclear Explosives in Cold War America*. Ithaca, NY: Cornell University Press.

Keith, David. 2013. *A Case for Climate Engineering*. Cambridge, MA: MIT Press.

Kelley, Robin D. G. 2003. *Freedom Dreams: The Black Radical Imagination*. Boston: Beacon.

Kelly, Kim. 2023. "The Young Miners Dying of 'An Old Man's Disease.'" *In These Times*, May 17. https://inthesetimes.com/article/coal-miners-black-lung-young-dying-old-mans-disease-kim-kelly.

Kenner, Alison. 2018. *Breathtaking: Asthma Care in a Time of Climate Change*. Minneapolis: University of Minnesota Press.

Kimball, Ward, dir. 1959. *Eyes in Outer Space: A Science Factual Presentation*. Walt Disney with the cooperation of the US Department of Defense. YouTube video, 25 min. https://www.youtube.com/watch?v=Xw9Leq98IRU.

King, Tiffany Lethabo. 2019. *The Black Shoals: Offshore Formations of Black and Native Studies*. Durham, NC: Duke University Press.

Kintisch, Eli. 2010. *Hack the Planet: Science's Best Hope—or Worst Nightmare—for Averting Climate Catastrophe*. Hoboken, NJ: Wiley.

Kleiman, Devra, and John Seidensticker. 1985. "Pandas in the Wild." *Science* 228 (4701): 875–76.

Kline, Ronald R. 2015. *The Cybernetics Moment: Or Why We Call Our Age the Information Age*. Baltimore: Johns Hopkins University Press.

Kolbert, Elizabeth. 2014. *The Sixth Extinction: An Unnatural History*. New York: Henry Holt.

Konsmo, Erin Marie, and Karyn Recollet. 2019. "Afterword: Meeting the Land(s) Where They Are At." In *Decolonizing Studies in Education: Mapping the Long View*, edited by Linda Tuhiwai Smith, Eve Tuck, and K. Wayne Yang, 238–51. New York: Routledge.

Kosek, Jake. 2006. *Understories: The Political Life of Forests in Northern New Mexico*. Durham, NC: Duke University Press.

Kosek, Jake. 2010. "Ecologies of Empire: On the New Uses of the Honeybee." *Cultural Anthropology* 25 (4): 650–78.

Kosek, Jake. Forthcoming. *Homo Apians*. Durham, NC: Duke University Press.

LaDuke, Winona. 2017. *All Our Relations: Native Struggles for Land and Life*. Chicago: Haymarket.

la paperson. 2017. *A Third University Is Possible*. Minneapolis: University of Minnesota Press.

Lartey, Jamiles. 2019. "'It's Totally Unfair': Chicago, Where the Rich Live 30 Years Longer Than the Poor." *Guardian*, June 23. https://www.theguardian.com/us -news/2019/jun/23/chicago-latest-news-life-expectancy-rich-poor-inequality.

Laville, Sandra. 2023. "Twelve Billionaires' Climate Emissions Outpollute 2.1M Homes, Analysis Finds." *Guardian*, November 19. https://www.theguardian .com/environment/2023/nov/20/twelve-billionaires-climate-emissions-jeff -bezos-bill-gates-elon-musk-carbon-divide.

Lawson, Alex, and Patrick Greenfield. 2023. "Shell to Spend $450m on Carbon Offsetting as Fears Grow That Credits May Be Worthless." *Guardian*, January 19. https://www.theguardian.com/environment/2023/jan/19/shell-to-spend -450m-on-carbon-offsetting-fears-grow-credits-worthless-aoe.

Le Quere, Corinne, Robert B. Jackson, Matthew W. Jones, et al. 2020. "Temporary Reduction in Daily Global CO_2 Emissions During the COVID-19 Forced Confinement." *Nature Climate Change* 10 (July): 647–53.

Lee, Jennifer 8. 2003. "A Call for Softer, Greener Language." *New York Times*, March 2. https://www.nytimes.com/2003/03/02/us/a-call-for-softer-greener -language.html.

Liang, Guanxiang, and Frederic D. Bushman. 2021. "The Human Virome: Assembly, Composition and Host Interactions." *Nature Reviews Microbiology* 19 (8): 514–27. https://doi.org/10.1038/s41579-021-00536-5.

Liboiron, Max. 2021. *Pollution Is Colonialism*. Durham, NC: Duke University Press.

Lindqvist, Sven. 2007. *Terra Nullius: A Journey Through No One's Land*. New York: New Press.

Lipsitz, George. 2006. *The Possessive Investment in Whiteness: How White People Profit from Identity Politics*. Rev. and expanded ed. Philadelphia: Temple University Press.

Livesay, Nora, and John D. Nichols, eds. 2021. *The Ojibwe People's Dictionary*. Minneapolis: University of Minnesota Libraries. https://ojibwe.lib.umn.edu/.

Lockett, Will. 2020. "Is Elon Musk Right? Should We Nuke Mars?" *Medium*, October 6. https://medium.com/predict/is-elon-musk-right-should-we-nuke -mars-1cec8b6093af.

Lorde, Audre. 2007. "The Uses of the Erotic: The Erotic as Power." In *Sister Outsider: Essays and Speeches*, 53–59. Berkeley, CA: Crossing Press.

Lorimer, Jamie. 2020. *The Probiotic Planet: Using Life to Manage Life*. Minneapolis: University of Minnesota Press.

Lowe, Lisa. 2015. *The Intimacies of Four Continents*. Durham, NC: Duke University Press.

Lyall, Sarah. 2009. "Fondly, Greenland Loosens Danish Rule." *New York Times*, June 21. https://www.nytimes.com/2009/06/22/world/europe/22greenland .html.

Lynas, Mark, Benjamin Z. Houlton, and Simon Perry. 2021. "Greater Than 99% Consensus on Human Caused Climate Change in the Peer-Reviewed Scientific Literature." *Environmental Research Letters* 16 (11): 144005.

Mahmood, Saba. 2011. *Politics of Piety: The Islamic Revival and the Feminist Subject*. Princeton, NJ: Princeton University Press.

Main, Thomas J. 2021. *The Rise of Illiberalism*. Washington, DC: Brookings Institution.

Marcott, Shaun, and Jeremy Shakun. 2021. "A Complete Palaeoclimate Picture Emerges." *Nature* 599 (7884): 208–9.

Martin, Laura J. 2018. "Proving Grounds: Ecological Fieldwork in the Pacific and the Materialization of Ecosystems." *Environmental History* 23 (3): 567–92.

Marx, Karl. (1867) 2004. *Capital: Volume I*. London: Penguin.

Marx, Karl, with Friedrich Engels. (1932) 1998. *The German Ideology*. Amherst, NY: Prometheus.

Masco, Joseph. 2014. *The Theater of Operations: National Security Affect from the Cold War to the War on Terror*. Durham, NC: Duke University Press.

Masco, Joseph. 2020. *The Nuclear Borderlands: The Manhattan Project in Post–Cold War New Mexico*. New ed. Princeton, NJ: Princeton University Press.

Masco, Joseph. 2021a. "The Artificial World." In *Reactivating Elements: Chemistry, Ecology, Practice*, edited by Dimitris Papadopoulos, María Puig de la Bellacasa, and Natasha Myers, 131–50. Durham, NC: Duke University Press.

Masco, Joseph. 2021b. *The Future of Fallout, and Other Episodes in Radioactive World-Making*. Durham, NC: Duke University Press.

Masco, Joseph, and Lisa Wedeen, eds. 2024. *Conspiracy/Theory*. Durham, NC: Duke University Press.

Materić, Dušan, Helle Astrid Kjær, Paul Vallelonga, Jean-Louis Tison, Thomas Röckmann, and Rupert Holzinger. 2022. "Nanoplastics Measurements in Northern and Southern Polar Ice." *Environmental Research* 208: 112741.

Mathews, Andrew S. 2022. *Trees Are Shape Shifters: How Cultivation, Climate Change, and Disaster Create Landscapes*. Durham, NC: Duke University Press.

Mbembe, Achille. 2003. "Necropolitics." *Public Culture* 15 (1): 11–40.

Mbembe, Achille. 2017. *Critique of Black Reason.* Durham, NC: Duke University Press.

McClintock, Anne. 1995. *Imperial Leather: Race, Gender, and Sexuality in the Colonial Context.* New York: Routledge.

McGregor, Deborah. 2009. "Honoring Our Relations: An Anishnaabe Perspective on Environmental Justice." In *Speaking for Ourselves: Environmental Justice in Canada*, edited by Julian Agyeman, Peter Cole, Randolph Haluza-DeLay, and Pat O'Riley, 27–41. Vancouver: UBC Press.

McIntosh, Emma. 2020. "Ontario Suspends Environmental Oversight Rules, Citing COVID-19." *Canada's National Observer*, April 8. https://www.national observer.com/2020/04/08/news/ontario-suspends-environmental -oversight-rules-citing-covid-19.

McKay, David I. A., Arie Staal, Jesse F. Abrams, et al. 2022. "Exceeding 1.5°C Global Warming Could Trigger Multiple Climate Tipping Points." *Science* 377 (6611). https://doi.org/10.1126/science.abn7950.

McKittrick, Katherine, ed. 2015. *Sylvia Wynter: On Being Human as Praxis.* Durham, NC: Duke University Press.

McKittrick, Katherine. 2021. *Dear Science and Other Stories.* Durham, NC: Duke University Press.

McNeill, John Robert, and Peter Engelke. 2014. *The Great Acceleration: An Environmental History of the Anthropocene Since 1945.* Cambridge, MA: Belknap.

Mehta, Uday Singh. 1999. *Liberalism and Empire: A Study in Nineteenth-Century British Liberal Thought.* Chicago: University of Chicago Press.

Messeri, Lisa. 2016. *Placing Outer Space: An Earthly Ethnography of Other Worlds.* Durham, NC: Duke University Press.

Metzl, Jonathan M. 2011. *The Protest Psychosis: How Schizophrenia Became a Black Disease.* Boston: Beacon.

Milbank, Dana, and Claudia Dean. 2003. "Hussein Link to 9/11 Lingers in Many Minds." *Washington Post*, September 6. https://www.washingtonpost.com /archive/politics/2003/09/06/hussein-link-to-911-lingers-in-many-minds /7cd31079-21d1-42cf-8651-b67e93350fde/.

Mills, Charles W. 1997. *The Racial Contract.* Ithaca, NY: Cornell University Press.

Mingus, Mia. 2019. "The Four Parts of Accountability and How to Give a Genuine Apology." *Leaving Evidence*, December 18. https://leavingevidence .wordpress.com/2019/12/18/how-to-give-a-good-apology-part-1-the-four -parts-of-accountability/.

Misrach, Richard, and Kate Orff. 2014. *Petrochemical America.* New York: Aperture.

Missing Migrants Project. 2023. "30,219 Missing Migrants Recorded in Mediterranean (Since 2014)." January 17. https://missingmigrants.iom.int/region /mediterranean.

Mitchell, Audra, and Aadita Chaudhury. 2020. "Worlding Beyond 'the End of the World': White Apocalyptic Vision and BIPOC Futurisms." *International Relations* 34 (3): 309–32.

Mitchell, P. I., L. León Vintró, H. Dahlgaard, C. Gascó, and J. A. Sánchez-Cabeza. 1997. "Perturbation in the ^{240}Pu ^{239}Pu Global Fallout Ratio in Local Sediments Following the Nuclear Accidents at Thule (Greenland) and Palomares (Spain)." *Science of the Total Environment* 202 (1): 147–53.

Moreton-Robinson, Aileen. 2015. *The White Possessive: Property, Power, and Indigenous Sovereignty*. Minneapolis: University of Minnesota Press.

Morrison, Toni. 1995. "The Site of Memory." In *Inventing the Truth: The Art and Craft of Memoir*, edited by William Zinsser, 83–102. New York: Houghton Mifflin.

Morton, Oliver. 2016. *The Planet Remade*. Princeton, NJ: Princeton University Press.

Morton, Timothy. 2013. *Hyperobjects: Philosophy and Ecology After the End of the World*. Minneapolis: University of Minnesota Press.

Muñoz, José Esteban. 2009. *Cruising Utopia: The Then and There of Queer Futurity*. New York: NYU Press.

Murphy, M. 2006. *Sick Building Syndrome and the Problem of Uncertainty: Environmental Politics, Technoscience, and Women Workers*. Durham, NC: Duke University Press.

Murphy, M. 2008. "Chemical Regimes of Living." *Environmental History* 13 (4): 695–703.

Murphy, M. 2017a. "Alterlife and Decolonial Chemical Relations." *Cultural Anthropology* 32 (4): 494–503.

Murphy, M. 2017b. *The Economization of Life*. Durham, NC: Duke University Press.

Murphy, M. 2021. "Reimagining Chemical, With and Against Technoscience." In *Reactivating Elements: Chemistry, Ecology, Practice*, edited by Dimitris Papadopoulos, María Puig de la Bellacasa, and Natasha Myers, 257–79. Durham, NC: Duke University Press.

Myers, Joshua. 2023. *Of Black Study*. London: Pluto.

National Academies of Sciences, Engineering, and Medicine. 2021. *Reflecting Sunlight: Recommendations for Solar Geoengineering Research and Research Governance*. Washington, DC: National Academies Press. https://doi.org /10.17226/25762.

National Research Council. 2015a. *Climate Intervention: Carbon Dioxide Removal and Reliable Sequestration*. Washington, DC: National Academies Press.

National Research Council. 2015b. *Climate Intervention: Reflecting Sunlight to Cool Earth*. Washington, DC: National Academies Press.

National Science Foundation. 1965. *Weather and Climate Modification: Report of the Special Commission on Weather Modification*. Washington, DC: US Government Printing Office.

Neocleous, Mark. 2008. *Critique of Security*. Edinburgh: Edinburgh University Press.

Ngai, Mae M. 2014. *Impossible Subjects: Illegal Aliens and the Making of Modern America*. Princeton, NJ: Princeton University Press.

Nichols, Robert. 2020. *Theft Is Property! Dispossession and Critical Theory*. Durham, NC: Duke University Press.

Nieder, Rolf, Dinesh K. Benbi, and Franz X. Reichl. 2018. "Soil-Borne Particles and Their Impact on Environment and Human Health." In *Soil Components and Human Health*. Dordrecht: Springer. https://doi.org/10.1007/978-94-024 -1222-2_3.

Oberg, James E. 1981. *New Earths: Transforming Other Planets for Humanity*. Mechanicsburg, PA: Stackpole.

Odum, Eugene P., and Edward J. Kuenzler. 1963. "Experimental Isolation of Food Chains in an Old-Field Ecosystem with the Use of Phosphorus-32." *Radioecology* 113: 113–21.

Ogle, Vanessa. 2015. *The Global Transformation of Time, 1870–1950*. Cambridge, MA: Harvard University Press.

Olson, Valerie. 2018. *Into the Extreme: U.S. Environmental Systems and Politics Beyond Earth*. Minneapolis: University of Minnesota Press.

Omi, Michael, and Howard Winant. 2015. *Racial Formation in the United States*. 3rd ed. New York: Routledge.

Omura, Keiichi, Grant Jun Otsuki, Shiho Satsuka, and Atsuro Morita. 2018. *The World Multiple: The Quotidian Politics of Knowing and Generating Entangled Worlds*. New York: Routledge.

O'Neill, Gerard K. 1976. *The High Frontier: Human Colonies in Space*. New York: William Morrow.

O'Neill, Gerard K., and Ginie Reynolds. 1977. "Habitats in Space." *Science Teacher* 44 (6): 22–26.

Oreskes, Naomi. 2004. "The Scientific Consensus on Climate Change." *Science* 306 (5706): 1686.

Oreskes, Naomi. 2021. *Science on a Mission: How Military Funding Shaped What We Do and Don't Know About the Ocean*. Chicago: University of Chicago Press.

Oreskes, Naomi, and Erik M. Conway. 2011. *Merchants of Doubt: How a Handful of Scientists Obscured the Truth on Issues from Tobacco Smoke to Global Warming*. New York: Bloomsbury.

Oreskes, Naomi, with Homer Le Grand, eds. 2003. *Plate Tectonics: An Insider's History of the Modern Theory of the Earth*. Boulder, CO: Westview Press.

Orr, Jackie. 2006. *Panic Diaries: A Genealogy of Panic Disorder*. Durham, NC: Duke University Press.

Osman, Matthew B., Jessica E. Tierney, Jiang Zhu, et al. 2021. "Globally Resolved Surface Temperature Since the Last Glacial Maximum." *Nature* 599: 239–44. https://doi.org/10.1038/s41586-021-03984-4.

Papadopoulos, Dimitris, María Puig de la Bellacasa, and Natasha Myers, eds. 2021. *Reactivating Elements: Chemistry, Ecology, Practice*. Durham, NC: Duke University Press.

Parikka, Jussi. 2015. *A Geology of Media*. Minneapolis: University of Minnesota Press.

Parreñas, Juno Salazar. 2018. *Decolonizing Extinction: The Work of Care in Orangutan Rehabilitation*. Durham, NC: Duke University Press.

Pasternak, Shiri. 2017. *Grounded Authority: The Algonquins of Barriere Lake Against the State*. Minneapolis: University of Minnesota Press.

Pasternak, Shiri, Hayden King, and Riley Yesno. 2019. *Land Back: A Yellowhead Institute Red Paper*. Toronto: Yellowhead Institute.

Pearce, Trevor. 2010. "From 'Circumstances' to 'Environment': Herbert Spencer and the Origins of the Idea of Organism–Environment Interaction." *Studies in History and Philosophy of Biological and Biomedical Sciences* 41 (3): 241–52.

Perkins, Tom. 2022. "'Forever Chemicals' Detected in All Umbilical Cord Blood in 40 Studies." *Guardian*, September 23. https://www.theguardian.com/environment/2022/sep/23/forever-chemicals-found-umbilical-cord-blood-samples-studies.

Pignarre, Phillipe, and Isabelle Stengers. 2011. *Capitalist Sorcery: Breaking the Spell*. New York: Palgrave Macmillan.

Pistor, Katharina. 2019. *The Code of Capital: How the Law Creates Wealth and Inequality*. Princeton, NJ: Princeton University Press.

Political Economy Research Institute. 2022. *Greenhouse 100 Polluters Index (2022 Report, Based on 2020 Data)*. Amherst: University of Massachusetts Amherst. https://peri.umass.edu/greenhouse-100-polluters-index-current.

Popovich, Nadja, and Brad Plumer. 2021. "Who Has the Most Historical Responsibility for Climate Change?" *New York Times*, November 12. https://www.nytimes.com/interactive/2021/11/12/climate/cop26-emissions-compensation.html.

Povinelli, Elizabeth A. 2001. "Radical Worlds: The Anthropology of Incommensurability and Inconceivability." *Annual Review of Anthropology* 30: 319–34.

Povinelli, Elizabeth A. 2016. *Geontologies: A Requiem to Late Liberalism*. Durham, NC: Duke University Press.

Pratt, Mary Louise. 1992. *Imperial Eyes: Travel Writing and Transculturation*. New York: Routledge.

Public Enemy. 1990. *Fear of a Black Planet*. New York: Def Jam Recordings.

Ra, Sun. 2011. *This Planet Is Doomed*. Brooklyn: Kicks Books.

Radio Canada International. 2018. "Soaring Warm Temperatures in Arctic." February 22. https://www.rcinet.ca/en/2018/02/22/soaring-warm-temperatures-in-arctic/.

Raffles, Hugh. 2010. "Air." In *Insectopedia*, 5–12. New York: Vintage.

Rankin, William. 2018. *After the Map: Cartography, Navigation, and the Transformation of Territory in the Twentieth Century*. Chicago: University of Chicago Press.

Reagon, Bernice Johnson. 1983. "Coalition Politics: Turning the Century." In *Home Girls: A Black Feminist Anthology*, edited by Barbara Smith, 356–68. New York: Kitchen Table (Women of Color Press).

Reddy, Elizabeth. 2014. "What Does It Mean to Do Anthropology in the Anthro-

pocene?" *Platypus: The CASTAC Blog*, April 8. https://blog.castac.org/2014/04
/what-does-it-mean-to-do-anthropology-in-the-anthropocene/.

Rettner, Rachael. 2018. "Meet Your Interstitium, a Newfound 'Organ.'" *Scientific American*, March 27. https://www.scientificamerican.com/article/meet
-your-interstitium-a-newfound-organ/.

Roberts, Elizabeth F. S. 2017. "What Gets Inside: Violent Entanglements and Toxic Boundaries in Mexico City." *Cultural Anthropology* 32 (4): 592–619.

Robinson, Cedric J. 2020. *Black Marxism: The Making of the Black Radical Tradition*. 3rd ed. Berkeley: University of California Press.

Robinson, Kim Stanley. 1993. *Red Mars*. New York: Spectra-Bantam.

Robinson, Kim Stanley. 1994. *Green Mars*. New York: Spectra-Bantam.

Robinson, Kim Stanley. 1996. *Blue Mars*. New York: Spectra-Bantam.

Robinson, Kim Stanley. 2020. *The Ministry for the Future*. New York: Orbit.

Rochman, Chelsea M., Cole Brookson, Jacqueline Bikker, et al. 2019. "Rethinking Microplastics as a Diverse Contaminant Suite." *Environmental Toxicology and Chemistry* 38 (4): 703–11. https://doi.org/10.1002/etc.4371.

Roediger, D. R. 2017. "The Wages of Whiteness: Race and the Making of the American Working Class." In *Class: The Anthology*, edited by Stanley Aronowitz and Michael J. Roberts, 41–55. Oxford: John Wiley and Sons.

Royal Society. 2009. *Geoengineering the Climate: Science, Governance and Uncertainty*. London: Royal Society.

Rudwick, Martin J. S. 1997. *Georges Cuvier, Fossil Bones, and Geological Catastrophes: New Translations and Interpretations of the Primary Texts*. Chicago: University of Chicago Press.

Rudwick, Martin J. S. 2014. *Earth's Deep History: How It Was Discovered and Why It Matters*. Chicago: University of Chicago Press.

Ruíz, Elena. 2020. "Cultural Gaslighting." *Hypatia* 35 (4): 687–713.

Schivelbusch, Wolfgang. 2014. *The Railway Journey: The Industrialization of Time and Space in the Nineteeth Century*. Berkeley: University of California Press.

Scranton, Roy. 2015. *Learning to Die in the Anthropocene: Reflections on the End of a Civilization*. San Francisco: City Lights.

Sedgwick, Eve Kosofsky. 2008. *Epistemology of the Closet*. Rev. ed. Berkeley: University of California Press.

Seligman, Lara. 2024. "Confused About Biden's Israel Weapons Policy? Here's What You Should Know." *Politico*, May 15. https://www.politico.com/news
/2024/05/15/biden-israel-weapons-policy-00158210.

Shadaan, Reena, and Michelle Murphy. 2020. "EDC's as Industrial Chemicals and Settler Colonial Structures: Towards a Decolonial Feminist Approach." *Catalyst: Feminism, Theory, Technoscience* 6 (1). https://doi.org/10.28968/cftt
.v6i1.32089.

Shapiro, Ariel. 2021. "America's Biggest Owner of Farmland Is Now Bill Gates." *Forbes*, January 14. https://www.forbes.com/sites/arielshapiro/2021/01/14
/americas-biggest-owner-of-farmland-is-now-bill-gates-bezos-turner/.

Shapiro, Nicholas. 2015. "Attuning to the Chemosphere: Domestic Formaldehyde, Bodily Reasoning, and the Chemical Sublime." *Cultural Anthropology* 30 (3): 368–93.

Sharpe, Christina. 2017. "Antiblack Weather vs. Black Microclimates." *Funambulist Magazine*, no. 14 (November 3). https://thefunambulist.net/magazine/14 -toxic-atmospheres/32058-2.

Shotwell, Alexis. 2016. *Against Purity: Living Ethically in Compromised Times*. Minneapolis: University of Minnesota Press.

Shotwell, Alexis. 2020. "The Virus Is a Relation." *Upping the Anti: A Journal of Theory and Action*, May 5. https://uppingtheanti.org/blog/entry/the-virus-is -a-relation.

Simmons, Kristen. 2017. "Settler Atmospherics." Member Voices, *Fieldsights*, November 20. https://culanth.org/fieldsights/settler-atmospherics.

Simpson, Leanne Betasamosake. 2014. "Land as Pedagogy: Nishnaabeg Intelligence and Rebellious Transformation." *Decolonization: Indigeneity, Education and Society* 3 (3): 1–25.

Sloterdijk, Peter. 2009a. "Airquakes." *Environment and Planning D: Society and Space* 27 (1): 41–57.

Sloterdijk, Peter. 2009b. *Terror from the Air*. Cambridge, MA: MIT Press.

Slotkin, Richard. 1998. *Gunfighter Nation: The Myth of the Frontier in Twentieth-Century America*. Norman: University of Oklahoma Press.

Slotkin, Richard. 2000. *Regeneration Through Violence: The Mythology of the American Frontier, 1600–1860*. Norman: University of Oklahoma Press.

Smith, Jen Rose. 2025. *Ice Geographies: The Colonial Politics of Race and Indigeneity in the Arctic*. Durham, NC: Duke University Press.

Smith, Linda Tuhiwai. 2012. *Decolonizing Methodologies: Research and Indigenous Peoples*. London: Zed.

Smith, Linda Tuhiwai, Eve Tuck, and K. Wayne Yang, eds. 2018b. *Indigenous and Decolonizing Studies in Education: Mapping the Long View*. New York: Routledge.

Smith, Shawn Michelle. 2004. *Photography on the Color Line: W. E. B. Du Bois, Race, and Visual Culture*. Durham, NC: Duke University Press.

Smith, Shawn Michelle. 2020. *Photographic Returns: Racial Justice and the Time of Photography*. Durham, NC: Duke University Press.

Soja, Edward W. 2013. *Seeking Spatial Justice*. Minneapolis: University of Minnesota Press.

Solar Geoengineering Non-Use Agreement. 2021. "Open Letter: We Call for an International Non-Use Agreement on Solar Geoengineering." https://www .solargeoeng.org/non-use-agreement/open-letter/.

Spencer, Herbert. 1864. *The Principles of Biology*. Vol. 1. London: Williams and Norgate.

Spivak, Gayatri Chakravorty. 2003. *Death of a Discipline*. New York: Columbia University Press.

Spivak, Gayatri Chakravorty. 2015. "'Planetarity' (Box 4, WELT)." *Paragraph* 38 (2): 290–92.

Starosielski, Nicole. 2015. *The Undersea Network*. Durham, NC: Duke University Press.

Steffen, Will, Katherine Richardson, Johan Rockstrom, et al. 2020. "The Emergence and Evolution of Earth System Science." *Nature Reviews Earth and Environment* 1: 54–63. https://doi-org/10.1038/s43017-019-0005-6.

Steffen, Will, Johan Rockström, Katherine Richardson, et al. 2018. "Trajectories of the Earth System in the Anthropocene." *PNAS* 115 (33): 8252–59.

Stengers, Isabelle. 2015. *In Catastrophic Times: Resisting the Coming Barbarism*. London: Open Humanities.

Stephenson, Neal. 2021. *Termination Shock*. New York: HarperCollins.

Sterling, Bruce. 1994. *Heavy Weather*. New York: Bantram Spectra.

Subcomandante Marcos. 2022. *Zapatista Stories for Dreaming An-Other World*. Translation, introduction and commentaries by Lightning Collective. Forward by JoAnne Wypijewski. Oakland, CA: PM Press.

Sultana, Farhana. 2022. "The Unbearable Heaviness of Climate Coloniality." *Political Geography* 99: 102638. https://doi.org/10.1016/j.polgeo.2022.102638.

Supran, G., S. Rahmstorf, and N. Oreskes. 2023. "Assessing ExxonMobil's Global Warming Projections." *Science* 379 (6628): eabk0063.

Suttle, Curtis A. 2005. "Viruses in the Sea." *Nature* 437 (15): 356–61.

Szwed, John. 2020. *Space Is the Place: The Lives and Times of Sun Ra*. Durham, NC: Duke University Press.

Taplin, Jonathan. 2017. *Move Fast and Break Things: How Facebook, Google, and Amazon Cornered Culture and Undermined Democracy*. New York: Little, Brown.

Technoscience Research Unit. 2019a. "The Land and the Refinery: Past, Present, and Future." Toronto: University of Toronto. https://www.landandrefinery.org/.

Technoscience Research Unit. 2019b. "Past, Present, Futures; Visualizing Chemical Valley Pollution and Colonialism Project Description." https://technoscienceunit.org/past-present-futures-visualizing-chemical-valley-pollution-and-colonialism-project-description/.

Ticktin, Miriam. 2017. "A World Without Innocence." *American Ethnologist* 44 (4): 577–90.

Todd, Zoe. 2017a. "Fish, Kin and Hope: Tending to Water Violations in *Amiskwaciwâskahikan* and Treaty Six Territory." *Afterall: A Journal of Art, Context and Enquiry* 43: 102–7.

Todd, Zoe. 2017b. "Protecting Life Below Water: Tending to Relationality and Expanding Oceanic Consciousness Beyond Coastal Zones." *American Anthropologist*, October 17. https://www.americananthropologist.org/deprovincializing-development-series/protecting-life-below-water.

Trapp, Robert J., Noah S. Diffenbaugh, Harold E. Brooks, Michael E. Baldwin, Eric D. Robinson, and Jeremy S. Pal. 2007. "Changes in Severe Thunderstorm

Environment Frequency During the 21st Century Caused by Anthropogen-
ically Enhanced Global Radiative Forcing." *Proceedings of the National Academy of Sciences* 104 (50): 19719–23.

Tsing, Anna Lowenhaupt. 2015. *The Mushroom at the End of the World: On the Possibility of Life in Capitalist Ruins*. Princeton, NJ: Princeton University Press.

Tsing, Anna Lowenhaupt, Nils Bubandt, Elaine Gan, and Heather Anne Swanson.
2017. *Arts of Living on a Damaged Planet: Ghosts and Monsters of the Anthropocene*. Minneapolis: University of Minnesota Press.

Tuck, Eve. 2009. "Suspending Damage: A Letter to Communities." *Harvard Educational Review* 79 (3): 409–27.

Tuck, Eve, and Marcia McKenzie. 2015. *Place in Research: Theory, Methodology, and
Methods*. New York: Routledge.

Tuck, Eve, and K. Wayne Yang. 2012. "Decolonization Is Not a Metaphor." *Decolonization: Indigeneity, Education and Society* 1 (1): 1–40.

Turchetti, Simone, and Peder Roberts, eds. 2014. *The Surveillance Imperative: Geosciences During the Cold War and Beyond*. New York: Palgrave Macmillan US.

Turner, Fred. 2013. *The Democratic Surround: Multimedia and American Liberalism
from World War II to the Psychedelic Sixties*. Chicago: University of Chicago
Press.

UNHCR. 2023. "Mid Year Trends 2023." United Nations High Commissioner for
Refugees. https://www.unhcr.org/mid-year-trends-report-2023.

United Nations. 2017. "Treaty on the Prohibition of Nuclear Weapons." https://
www.un.org/disarmament/wmd/nuclear/tpnw/.

United Nations. 2023. "Over 114 Million Displaced by War, Violence Worldwide."
UN News, October 25. https://news.un.org/en/story/2023/10/1142827#:~:text
=More%20than%20114%20million%20people,agency%20UNHCR%20said
%20on%20Wednesday.

United Nations Environment. 2023. "Scientific Assessment of the Ozone Layer
Depletion: 2022." *UNEP–UN Environment Programme*, January 9. http://www
.unep.org/resources/publication/scientific-assessment-ozone-layer
-depletion-2022.

United Nations Report. 2019. "Nature's Dangerous Decline 'Unprecedented';
Species Extinction Rates 'Accelerating.'" *United Nations Sustainable Development*, May 6. https://www.un.org/sustainabledevelopment/blog/2019/05
/nature-decline-unprecedented-report.

US Army. 2005. *Psychological Operations (Field Manual No. 3–05.30)*. Washington,
DC: Department of the Army.

Van Dooren, Thom. 2016. *Flight Ways: Life and Loss at the Edge of Extinction*. New
York: Columbia University Press.

Van Dooren, Thom. 2019. *The Wake of Crows: Living and Dying in Shared Worlds*.
New York: Columbia University Press.

Vertesi, Janet. 2015. *Seeing Like a Rover: How Robots, Teams, and Images Craft
Knowledge of Mars*. Chicago: University of Chicago Press.

Vimalassery, Manu. 2014. "Counter-Sovereignty." *J19: The Journal of Nineteenth-Century Americanists* 2 (1): 142–48.

Vimalassery, Manu, Juliana Hu Pegues, and Alyosha Goldstein. 2016. "Introduction: On Colonial Unknowing." *Theory and Event* 19 (4). https://muse.jhu.edu/article/633283.

Volcovici, Valerie. 2020. "Interior Department Seeks to Expedite Energy Projects to Speed COVID Recovery: Document." Reuters, September 2. https://www.reuters.com/article/us-health-coronavirus-environment-idUKKBN25T2JS.

Von Neumann, John. 1955. "Can We Survive Technology?" *Fortune*, June 1. https://fortune.com/2013/01/13/can-we-survive-technology/.

Voyles, Traci. 2015. *Wastelanding: Legacies of Uranium Mining in Navajo Country*. Minneapolis: University of Minnesota Press.

Walcott, Rinaldo. 2021. *The Long Emancipation: Moving Toward Black Freedom*. Durham, NC: Duke University Press.

Wald, Priscilla. 2008. *Contagious: Cultures, Carriers, and the Outbreak Narrative*. Durham, NC: Duke University Press.

Wall, Mike. 2019. "Looks Like Elon Musk Is Serious About Nuking Mars." *Space.com*, August 21. https://www.space.com/elon-musk-serious-nuke-mars-terraforming.html.

Wang, Y., H. Okochi, Y. Tani, et al. 2023. "Airborne Hydrophilic Microplastics in Cloud Water at High Altitudes and Their Role in Cloud Formation." *Environmental Chemistry Letters* 21: 3055–62. https://doi.org/10.1007/s10311-023-01626-x.

Wang, Zhuofeng, Jiaqi Zhang, Yong Yang, Mao Cao, Jiazi Ma, Shumin Li, Hua Shao, and Zhongjun Du. 2024. "Current Status, Trends, and Prediction in the Burden of Coal Workers' Pneumoconiosis in 204 Countries and Territories from 1990 to 2019." *Heliyon* 10 (19): e37940.

Warde, Paul, Libby Robin, and Sverker Sörlin, eds. 2018. *The Environment: A History of the Idea*. Baltimore: Johns Hopkins University Press.

Warner, Michael. 1993. *Fear of a Queer Planet: Queer Politics and Social Theory*. Minneapolis: University of Minnesota Press.

Waters, Colin N., and Simon D. Turner. 2022. "Defining the Onset of the Anthropocene." *Science* 378 (6621): 706–8.

Waters, Colin N., Jan Zalasiewicz, Colin Summerhayes, et al. 2016. "The Anthropocene Is Functionally and Stratigraphically Distinct from the Holocene." *Science* 351 (6269). https://doi.org/10.1126/science.aad2622.

Watts, Vanessa. 2013. "Indigenous Place-Thought and Agency Amongst Humans and Non-Humans (First Woman and Sky Woman Go on a European World Tour!)." *Decolonization: Indigeneity, Education and Society* 2 (1): 20–34.

Weizman, Eyal. 2017. *Forensic Architecture: Violence at the Threshold of Detectability*. New York: Zone.

Wekker, Gloria. 2016. *White Innocence: Paradoxes of Colonialism and Race*. Durham, NC: Duke University Press.

Wesling, Frances Cress. 1974. "The Cress Theory of Color-Confrontation and Racism." *Black Scholar* 5 (8): 32–40.

Westervelt, Amy. 2023. "Fossil Fuel Companies Donated $700M to US Universities over 10 Years." *Guardian*, March 1. https://www.theguardian.com/environment/2023/mar/01/fossil-fuel-companies-donate-millions-us-universities.

Weston, Kath. 2022. "Bequeathing a World: Ecological Inheritance, Generational Conflict, and Dispossession." *Cambridge Journal of Anthropology* 40 (2): 106–23.

Wezeman, Pieter D., Alexandra Kuimova, and Siemon T. Wezeman. 2022. "Trends in International Arms Transfers 2021." *SIPRI Fact Sheet*, March. https://www.sipri.org/sites/default/files/2022-03/fs_2203_at_2021.pdf.

White, Richard. 2011. *Railroaded: The Transcontinentals and the Making of Modern America*. New York: W. W. Norton.

Whitford, Emma. 2022. "College Endowments Boomed in Fiscal 2021." *Inside Higher Ed*, February 18. https://www.insidehighered.com/news/2022/02/18/college-endowments-boomed-fiscal-year-2021-study-shows.

Whyte, Kyle. 2017. "Indigenous Climate Change Studies: Indigenizing Futures, Decolonizing the Anthropocene." *English Language Notes* 55 (1–2): 153–62.

Whyte, Kyle. 2018. "Settler Colonialism, Ecology, and Environmental Injustice." *Environment and Society* 9 (1): 125–44. https://doi.org/10.3167/ares.2018.090109.

Wickberg, Adam, and Johan Gärdebo. 2022. *Environing Media*. London: Routledge.

Williams, Eric. (1944) 2021. *Capitalism and Slavery*. Chapel Hill: University of North Carolina Press.

WMO. 2022. "Executive Summary." In *Scientific Assessment of Ozone Depletion: 2022, GAW Report No. 278*. Geneva: World Meteorological Organization.

Women's Earth Alliance and Native Youth Sexual Health Network. 2016. *Violence on the Land, Violence on Our Bodies: Building an Indigenous Response to Environmental Violence*. http://landbodydefense.org/uploads/files/VLVBReportToolkit2016.pdf.

World Health Organization. 2014. "7 Million Premature Deaths Annually Linked to Air Pollution." March 25. https://www.who.int/news/item/25-03-2014-7-million-premature-deaths-annually-linked-to-air-pollution.

Wynter, Sylvia. 2003. "Unsettling the Coloniality of Being/Power/Truth/Freedom: Towards the Human, After Man, Its Overrepresentation—an Argument." *CR: The New Centennial Review* 3 (3): 257–337.

Yacovone, Donald. 2022. *Teaching White Supremacy: America's Democratic Ordeal and the Forging of Our National Identity*. New York: Pantheon.

Yusoff, Kathryn. 2018. *A Billion Black Anthropocenes or None*. Minneapolis: University of Minnesota Press.

Zalasiewicz, Jan, Colin N. Waters, Juliana A. Ivar do Sul, et al. 2016. "The Geological Cycle of Plastics and Their Use as a Stratigraphic Indicator of the Anthropocene." *Anthropocene* 13: 4–17.

Zalasiewicz, Jan, Colin N. Waters, Mark Williams, and Colin P. Summerhayes. 2019. *The Anthropocene as a Geological Time Unit: A Guide to the Scientific Evidence and Current Debate.* Cambridge: Cambridge University Press.

Zalasiewicz, Jan, Mark Williams, Colin N. Waters, Anthony Barnosky, and Peter Haff. 2014. "The Technofossil Record of Humans." *Anthropocene Review* 1 (1): 3–7.

Zee, Jerry C. 2022. *Continent in Dust: Experiments in a Chinese Weather System.* Berkeley: University of California Press.

Zimmer, Carl. 2015. *A Planet of Viruses.* Chicago: University of Chicago Press.

INDEX

Page numbers in italics indicate figures.

academic institutions in North America: COVID-19 pandemic and, 105–6; critical thinking and unlearning in, 30–33, 108, 110–12; environmental studies in, 16–17, 24–26, 30–32, 74–75; future of, 34–35, 140–41; gaslighting in, 110–11; as sites of resistance and transformation, 63, 78–79; terraformatics and, 124; violence embedded in, 11, 20–21, 25–26, 33, 43; Whiteness in, 30, 57–58, 59–61, 68–71. *See also* academic research and study

academic research and study: causality and, 135; desire-based, 134–135; disciplines in, 9–10, 20, 25–26, 74–75, 145; environmental crisis and, 1–2, 12, 15–16; limitations of, 81–82; objectification in, 23, 38–39, 51, 82–84; objectivity and, 7–8, 10, 38–39, 71, 131. *See also* academic institutions in North America; terraformatics

air and atmosphere, 95–98

air pollution. *See* pollution

alterlife, 147–48

Anishinaabe Land and Land relations, 89, 93–94

Anthropocene: ice cores and, 39–45, 47–48; interest in across scholarly disciplines, 17–20, 28–29; as term, *19*; planetary narratives and, 9–11, 39–41, 44, 47–48

anthropogenic extinction. *See* extinction of species

antiracist and postcolonial scholarship, 34–35, 66–67

anxiety. *See* fear and anxiety of White Supremacy

apocalypse narratives. *See* end-of-the-world narratives

appropriation of culture and ideas, 121, 126

Arendt, Hannah, 7n4

Bacon, Francis, 23

Barthes, Roland, 46

Batygin, Konstantin, 5

Bezos, Jeff, 26, *27*, 28

biosecurity, 104

Black studies. *See* antiracist and postcolonial scholarship

Bong Joon Ho, 13

breathing, 95–98, 101, 144

Brown, Michael, 5

Bush, George W., 57n3

Canada, colonization of, 88–92

Charismatic Mega Concepts (CMC), 41, 80–81, 116, 118. *See also* Anthropocene

chemistry discipline, 110–11

Christianity, 23, 65

citation, 126

classification and taxonomies, 50–*51*, 53

climate change, 9–10, *19*, 47, 57n3, 91, 111
CMC. *See* Charismatic Mega Concepts (CMC)
coal. *See* fossil fuel extraction; mining and minerals
collective learning. *See* joined-up work and study
collective work. *See* joined-up work and study
colleges. *See* academic institutions in North America
colonization: of outer space, 8–10, 26–28, 117; of what is now called Australia, 58; of what is now called Canada, 88–92
Columbus Telescope, 5–6, 6n3
conditions: of breathing, 95–98; instead of causality, 135; explanation of, 16–17n24, 22–23, 120–22; locating oneself in, 30–31, 131–32, 148–49; shifting of in your immediate relations, 125–26, 136–37; study of, 122
Cook, James, 58, 63
core sample, 39, *40*, 43–45, 47
coronavirus. *See* COVID-19 pandemic
COVID-19 pandemic, 101–6; academic institutions and, 105–6; health inequalities in the United States and, 101–4; racism and, 101–4; relationality and, 101, 104–5
cowriting, 32
Cress Wesling, Frances, 63n15
cultural appropriation. *See* appropriation of culture and ideas
Cuvier, Georges, 53

de la Cadena, Marisol, 122
democracy in the United States, 107
desire-based research, 134–35
despair, 33–34, 112
despondency. *See* despair
disinformation, 106–10, 112, 132
diversity in ways of being. *See* world of many worlds
Du Bois, W. E. B., 66

earth system, 24, *25*, 36n49, 39–43. *See also* environment, concept of
ecological crisis. *See* environmental crisis
ecology, 24. *See also* environment, concept of
Ediacaran Period, *14*
Ejército Zapatista de Liberación Nacional (EZLN). *See* Zapatistas
empiricism, 4, 7–8, 10, 21, 23, 50–51
end-of-the-world narratives, 55–56, 79, 139–41, 143. *See also* environmental crisis
Engineered Worlds conferences and seminars, 31, 145–51
environment, concept of, 16–17, 22–25, 58–60, 121, 145–56. *See also* conditions; earth system
environmental crisis: academic research and study and, 1–2, 15–16, 112; extinction of species and, 51–52; in science fiction, 13, 28n42; technocapitalist responses to, 26–29; White Supremacy and, 55–63, 73–74. *See also* end-of-the-world narratives
environmental regulations, 68–71, 102
environmental studies, 21–22, 24–26, 30–32, 74–75
extinction of species, 48–54; digital preservation and, 48–50; loss and, 51–54; species concept and, 50–53
EZLN. *See* Zapatistas

failure and mistakes, 77, 133–34
Fanon, Frantz, 34, 66–67n20, 76
FDWP. *See* Fear of a Dead White Planet (FDWP)
fear and anxiety of White Supremacy, 61–62, 64. *See also* Fear of a Dead White Planet (FDWP)
Fear of a Black Planet (Public Enemy), 63
Fear of a Dead White Planet (FDWP): death or undoing of, 73–79, 88, 127; explanation of, 2, 59–66; liberal and illiberal modes of, 68–73; past scholarship on, 66–67

fossil fuel extraction, 21, 44–45, 69, 71, 89–92
Foucault, Michel, 121

gaslighting, 108–11, 132
Gates, Bill, 26–28
geoengineering, 26–29, 71–72, 84–85, 117, 141
global warming. *See* climate change
golden spike: in geological science, 12–13, *14*; of the transcontinental railroad, 12–13, *14*
Great Exhibition (1851), 89, *90*
Greenland. *See* Kalaallit Nunaat

harm reduction, 81–82, 119, 123–24
health inequalities in the United States, 101–4
hopelessness. *See* despair
human exceptionalism, 52–53

ice cores, 39–45, 47–48
illiberalism, 68–71, 73, 102–3
imperial time. *See* time and linearity
Indigenous peoples: land dispossession and, 58, 79n37, 88–92, 109; relationships with land and, 58, 93–94, 116. *See also* Anishinaabe Land and Land relations; Métis Land relations; Zapatistas
interdependency. *See* relationality
Intergalactic Bummer Train: explanation of, 11–15; destination of, 77–78; breaking up of, 112–14
interstitium, 82–83

joined-up work and study: challenges of, 119, 125–26; examples of, 30–35, 145–56; necessity of, 111–13, 143–44; outside academic institutions, 132, 133; as a starting point, 77–78

Kalaallit Nunaat, 39, *40*, 42, 47
Karuka, Manu, 13, 57n5, 91n6
Kaua'i 'ō'ō, 48, *49*
kinship. *See* relationality

knowledge mastery. *See* mastery: of knowledge in one's discipline
Kuiper Belt, 4, *6*

land: colonization and dispossession of, 58, 88–92, 109; decolonization of, 47, 92–95; Indigenous peoples' relationships with, 58, 93–94, 116; as property or territory, 27n38, 88–93, 94–95; as relational, 92–94; as a source of language, knowledge, and law, 93–94; terraformatics and, 80, 86, 116. *See also* place-based modes of being
liberalism, 70–73, 102–3, 105
Liboiron, Max, 45–46, 93n13
linearity. *See* time and linearity
lungs, 95–98, 101

mapping and surveying, 88–91, 123n5
mastery: of knowledge in one's discipline, 71, 77, 81, 132, 133; over nature or the environment, 23, 50–51, 65. *See also* unknowing and unknowability of the world
media manipulation, 106–111, 132
methods: of classification, 50–51; of terraformatics, 37, 124–137; of terraforming better worlds, 37, 78–86; what is an X? 36, 38–39, 87–88
Métis Land relations, 93–94
microplastics. *See* plastics
middles, 76, 87–88, 93, 113, 136–37, 146
military, 22n31, 41–43, 47, 104
mining and minerals, 43, 48, 50
mistakes. *See* failure and mistakes
Moreton-Robinson, Aileen, 115
multiplicity in ways of being. *See* world of many worlds
Murphy, M., 147–48
Musk, Elon, 26, *27*, 28
mutual aid, 105

NASA, 8, 117
nature. *See* environment, concept of; mastery: over nature or the environment

nuclear nationalism, 69–70
nuclear warfare and technology, 42–43, 47

objectification: in scientific research, 23,
    38–39, 51, 82–84; Whiteness and, 23, 58,
    60, 64
objectivity and object-centered research,
    7–8, 10, 38–39, 71, 131
oil. *See* fossil fuel extraction
One Worldism: Charismatic Mega Con-
    cepts and, 80; earth system and, 36n49;
    explanation of, 2, 10; terraformatics and,
    117, 122–23, 129–30; White Supremacy
    and, 64–66. *See also* planetary narra-
    tives; world of many worlds
organs in the human body, 82–83
outer space colonization. *See* coloniza-
    tion: of outer space

performance of mastery. *See* mastery: of
    knowledge in one's discipline
petroleum. *See* fossil fuel extraction
place-based modes of being, 47–48,
    92–94, 116, 125. *See also* land
planetary emergency. *See* environmental
    crisis
planetary narratives: in academic insti-
    tutions and research, 9–11; in ice cores
    and the Anthropocene, 9–11, 39–41,
    44, 47–48; microplastics and, 44–46;
    nuclear and military technology and,
    42–43; One Worldism and, 65–66; revo-
    lutionary potential in, 66n19. *See also*
    One Worldism
Planet Nine, 4–6, 8, 10–11
plastics, 44–46, 48, 71
pluralism. *See* world of many worlds
pollution, 92, 96–98
postcolonial studies. *See* antiracist and
    postcolonial scholarship
propaganda, 107–9, 132
psyops (psychological operations),
    107–10, 112. *See also* disinformation;
    gaslighting; propaganda
Public Enemy, 63

public health in the United States, 102–4

racism and the COVID-19 pandemic, 101–4.
    See also Whiteness; White Supremacy
railroad. *See* transcontinental railroad
Red River Rebellion, 90
relationality: of animals with other beings,
    48, 53–54; of breathing, 98; instead of
    causality, 135; COVID-19 pandemic and,
    101, 104–5; of land, 92–94; of micro-
    plastics, 45–46; of organs in the human
    body, 82–83; of subjects of study, 38–39
respiration. *See* breathing
Robinson, Cedric, 79n37

SARS-CoV-2. *See* COVID-19 pandemic
savior narratives, 17, 52, 71–72, 119
scholarly research. *See* academic research
    and study
science fiction, 13, 28n42, 85n47, 117
scientific mastery. *See* mastery
scientific research. *See* academic research
    and study
server farms, 48–50
"shareholder whiteness," 13, 57n5. *See also*
    Whiteness
Snowpiercer (Bong Joon Ho), 13
social constructionism. *See* conditions
solidarity. *See* joined-up work and study
solutionism and resistance of, 14, 71,
    79–80, 128–29, 138–39
species concept, 50–54, 99
species extinction. *See* extinction of
    species
Spencer, Herbert, 16
Spivak, Gayatri Chakravorti, 66n19
studying differently. *See* terraformatics
surveying land. *See* mapping and
    surveying

taxonomies. *See* classification and
    taxonomies
technocapitalist solutions, 26–28, 105
terraformatics: explanation of, 36,
    116–19, 122–24; land- and place-based

study and, 80, 86, 116; limitations of, 81–82; methods of, 124–37. *See also* terraforming

terraformations. *See* terraforming

terraforming: as dismantling, breaking, undoing, and refusing violent concepts, practices, and objects, 75–77, 88, 127, 132–33; explanation of, 36, 120, 149–50; of more habitable and less hostile worlds, 78–86; in science fiction, 85n47, 117; viruses as, 99–101; White Supremacy as, 58–60. *See also* terraformatics

terra nullius, 27n38, 58, 109

time and linearity, 13, 39, 130–31

Todd, Zoe, 44, 45n63

trains. *See* Intergalactic Bummer Train; transcontinental railroad

transcontinental railroad, 12–13, *14*

triangulation, 110–11

Trump administration, 57n3, 69n27

universalism. *See* One Worldism

universal time. *See* time and linearity

universities. *See* academic institutions in North America

unknowing and unknowability of the world, 4–6, 129–30, 136. *See also* mastery: of knowledge in one's discipline

unlearning, 30–31, 34–35, 108

viruses, 99–*100*

voting. *See* democracy in the United States

water pollution. *See* pollution

weather forecasting, 42, 72n2

what is an X?: applied to ice cores and worlds, 39–48; applied to land, 88–95; applied to lungs, 95–98; as a method, 36, 38–39, 87–88, 136; applied to microplastics, 45–46; applied to species and loss, 48–54; applied to thinking, 106–14; applied to viruses, 99–106

Whiteness: in academic institutions in North America, 30, 57–58, 59–61, 68–71; explanation of, 57–58, 65; hierarchies and, 65; innocence and, 131–132; objectification and, 23, 58, 60, 64. *See also* "shareholder whiteness"; White Supremacy

White possession, 57–58, 64, 71, 90, 115

White savior narratives. *See* savior narratives

White Supremacy: atmosphere as a weapon of, 97–98; Cress Wesling's theory of, 63n15; environmental crisis and, 55–63, 73–74; fear and anxiety and, 61–62, 64; liberal and illiberal modes of, 68–73; One Worldism and, 64–66. *See also* Whiteness; White possession

worlding. *See* terraforming

world of many worlds: academic institutions and, 124, 129; explanation of, 2, 12, 122–23; land and, 27n38, 47–48; unknowing and unknowability of the world and, 136. *See also* One Worldism

Zapatistas, 2n1, 122, 136

www.ingramcontent.com/pod-product-compliance
Lightning Source LLC
Chambersburg PA
CBHW020534270326
41927CB00006B/576